Biotechnology Applications of Microinjection, Microscopic Imaging, and Fluorescence

Biotechnology Applications of Microinjection, Microscopic Imaging, and Fluorescence

Edited by

Peter H. Bach
C. Hugh Reynolds
Jessica M. Clark
John Mottley
Phil L. Poole

University of East London
London, United Kingdom

Plenum Press • New York and London

Library of Congress Cataloging-in-Publication Data

Biotechnology applications of microinjection, microscopic imaging, and
  fluorescence / edited by Peter H. Bach ... [et al.].
        p.   cm.
    "Proceedings of the First European Workshop on Microscopic
  Imaging, Fluorescence, and Microinjection in Biotechnology, held
  April 21-24, 1992, in London, United Kingdom"--T.p. verso.
    Includes bibliographical references and index.
    ISBN 0-306-44497-6
    1. Microinjections--Congresses. 2. Confocal microscopy-
  -Congresses. 3. Fluorescence spectroscopy--Congresses.
  4. Molecular probes--Congresses. 5. Biotechnology--Methodology-
  -Congresses.   I. Bach, P. H. (Peter H.) II. European Workshop on
  Microscopic Imaging, Fluorescence, and Microinjection in
  Biotechnology (1st : 1992 : London, England)
  TP248.24.B563   1993
  610'.72--dc20   93-10726
                                                             CIP

Proceedings of the First European Workshop on Microscopic Imaging, Fluorescence, and Microinjection in Biotechnology, held April 21–24, 1992, in London, United Kingdom

ISBN 0-306-44497-6

©1993 Plenum Press, New York
A Division of Plenum Publishing Corporation
233 Spring Street, New York, N.Y. 10013

Printed in the United States of America

PREFACE

Individual cells behave in surprising ways that cannot be deduced from the averaged results of an organ as assessed by the use of conventional biochemical methods. Thus multicellular plant and animals systems are being investigated by an increasing array of histochemical and cytochemical techniques based on general chemical or specific immunological interactions to identify structural materials and to assess biological activities.

In recent years there has been an increasing range of fluorescent probes, along with advanced computerised imaging and analysis techniques, which allows the behaviour of individual living cells to be followed in considerable detail. The parallel use of microinjection, microelectrodes and patch-clamping provides additional information about cells and their responses. Recombinant DNA technology has highlighted the desirability and the power of microinjecting defined materials into specific cells and so manipulating their fundamental biochemistry. New hypotheses are being tested which will form the cornerstone of future developments across the whole spectrum of biotechnology.

The *First European Workshop on Biotechnology Applications of Microinjection, Microscopic Imaging and Fluorescence* was run at the University of East London, U.K, 21st-24th April, 1992 with the objective of bringing together a diverse group of individuals who were using these state-of-the-art applications for biotechnological exploration. A novel feature of the meeting was participation by instrument manufacturers in the programme: there were hands-on workshops (where living cells could be examined), combined with the poster sessions.

This meeting attracted cell biologists of all persuasions, instrument scientists, physical chemists and information scientists who shared an interest in a series of new concepts. These included correlating the physicochemical properties of fluorescent probe molecules with their distribution in cells; the use of synthetic fluorescent lipid molecules to study lipid dynamics; the exploitation of the fluorescence of natural molecules; the full range of microinjection and patch-clamping applications to turn "the living cell into a test-tube" and investigate signal transduction, intracellular communication and development; all aspects of microscopic imaging and its applications in physiology, pharmacology and toxicology, especially using confocal laser scanning microscopy to facilitate determination of the three-dimensional distribution of fluorescence. Other topics included flow cytometry, neuronal organisation in the human brain, single-cell laser microsurgery, chromatin folding and low-light-intensity imaging and image analysis.

Many of the techniques are sophisticated, and much of the equipment is expensive; thus research based on these powerful techniques can best be advanced by laboratory exchanges, something that can now take place in Europe under a number of European Commission programmes.

The meeting was supported in part by the European Commission "Bridge" Biotechnology Programme, and also by several of the instrument manufactures. The workshop was greatly facilitated by the efforts of Joanne Thorndike, Stephen Brant, Tony Anthonypillai, Ambi Ambihaipahan, Sarita Mohur, Tom McCartney, Steve Wall and the technical staff who made the hands-on part of the workshop possible.

Faculty of Science   Peter Bach
University of East London   Hugh Reynolds
Romford Road   Jessica Clark
London, United Kingdom   John Mottley and
                                                                     Phil Poole

April 1993

ACKNOWLEDGEMENTS

This Workshop was supported in part by the Commission of the European Communities' "BRIDGE" Biotechnology Programme.

Exhibitors present

Carl Zeiss (Oberkochen) Ltd
Tracor Europa
Micro Instruments Ltd
Olympus Optical Co. (UK) Limited
Nikon (UK) Limited
Applied Imaging International Ltd
Braun Biotech
Biotech Instruments Ltd
Bio-Rad Microscience Ltd
Hamamatsu Photonics (UK) Ltd
Narishige Europe Ltd
Bachofer, Germany
Newcastle Photometrics Systems

CONTENTS

Contents

MICROINJECTION: A TECHNIQUE TO STUDY INHIBITION OF HIV-1 REPLICATION MEDIATED BY ANTISENSE RNA AND PARVOVIRUS GENES

Georg Sczakiel, Ingo Oelze, and Karola Rittner

Deutsches Krebsforschungszentrum, Angewandte Tumor-virologie (ATV), Im Neuenheimer Feld 242, D-6900 Heidelberg, Germany

INTRODUCTION

Infection with human immunodeficiency viruses is the causative event for development of the acquired immunodeficiency syndrome (AIDS; Barre-Sinoussi *et al.*, 1983; Popovic *et al.*, 1984). The search for and the testing of inhibitors of the human immunodeficiency virus type 1 (HIV-1) with tissue culture cells depends on appropriate assay systems for applying such substances and quantifying HIV-1 replication and gene expression. Further, it is of great interest to directly compare the antiviral effects of chemically different molecules such as proteins, short or large nucleic acids and low molecular weight compounds. From the technical viewpoint, capillary microinjection enables the delivery of water-soluble substances of any chemical nature directly into cells. For example, successful capillary microinjection experiments have been performed with very large particles such as intact coxsackie B1 viruses (Modalsli *et al.*, 1991) or yeast YAC DNA (660 kb; Gnirke and Huxley, 1991) as well as with antibodies (Riabowol *et al.*, 1988; LaMorte *et al.*, 1992) or other proteins (Smith *et al.*, 1989) and the relatively small deoxynucleotides (Wawra, 1988).

Microinjection of infectious proviral HIV-1 DNA into the nucleus of various vertebrate fibroblast cell lines leads to the production and release of infectious virus particles (Boyd *et al.*, 1988). Based on this finding, we established a quantifiable assay for HIV-1 replication which was used to study inhibition of HIV-1 mediated by antisense RNA (Sczakiel *et al.*, 1990; Rittner and Sczakiel, 1991) and by the *rep* gene of the adeno-associated virus type 2 (AAV-2; Rittner *et al.*, 1991a).

Biotechnology Applications of Microinjection, Microscopic Imaging, and Fluorescence, Edited by P.H. Bach *et al.*, Plenum Press, New York, 1993

1

MATERIALS AND METHODS

Cell Lines

The human epitheloid cell line SW480 (Leibovitz *et al.*, 1976) was used for microinjection experiments. SW480 cells were maintained at 37°C/5% CO_2 in DMEM (Biochrom, Berlin) supplemented with 10% foetal calf serum, L-glutamine (2 mM), penicillin (100 U/ml) and streptomycin (100 mg/ml). The human T-lymphoid cell line MT-4 (Harada *et al.*, 1985) was used for amplification of HIV-1 in co-cultures with SW480 cells. MT-4 cells were grown in RPMI 1640 (Biochrom, Berlin) at 37°C and 5% CO_2 with the supplements described above. For microinjection SW480 cells were grown on cover slides.

Plasmids and Synthetic RNA

The infectious proviral HIV-1 clone pNL43 has been described elsewhere (Adachi *et al.*, 1986). Expression plasmids for T7 promoter-driven *in vitro* transcription and for intracellular expression of HIV-1-directed antisense RNA have been described in detail by Sczakiel *et al.* (1990) and by Rittner and Sczakiel (1991). Cloned AAV-2 genomes have been described by Heilbronn *et al.* (1990). The eucaryotic expression plasmid pKEX which was used for expression of antisense RNA, the AAV-2 *rep* gene and the chloramphenicol acetyltransferase gene has been described by Rittner *et al.* (1991a). Briefly, transcription from the eucaryotic expression cassette is driven by the strong and constitutive immediate early promoter/enhancer element of the human cytomegalovirus and terminated by the SV40 t-splice and polyadenylation signals.

Preparation of Samples for Microinjection

All plasmids used in microinjection experiments were purified by one CsCl gradient centrifugation. Since traces of CsCl contained in microinjected solutions can harm microinjected cells, plasmid DNAs were further purified by ion exchange chromatography with DE-52 cellulose (Whatman, UK) or prepacked commercial "Quiagen" columns (Diagen, Germany) using a binding buffer of 250 mM ammonium acetate, 5 mM $MgCl_2$, 0.5 mM EDTA, 0.05% SDS; a washing buffer of 10 mM Tris/HCl pH 8.0, 1 mM EDTA and an elution buffer of 2 M NaCl. The concentrations of nucleic acids were determined by UV-spectroscopy and the composition and concentrations of components of mixtures prepared for microinjection were controlled by agarose gel electrophoresis with linearized plasmid DNAs.

Preparation of Glass Capillaries for Microinjection

Glass capillaries (Clark Electromedical Instruments, UK, GC150-10) were prepared to a diameter of approximately 1 mm by the manual pipette puller MPP1 (Brindi, Germany) with two separate strokes. 1st stroke: elasticity 4.5; brake 35 mm; current 5.8 A. 2nd stroke: elasticity 12.0; no brake; current 5.0 A. The listed parameters depend in parts on the geometry of the heated platinum/iridium coil and have to be optimized for each individual puller or coil respectively.

Microinjection of Human Epitheloid Cells

Capillaries were filled with test solutions through the rear opening by means of another glass capillary drawn to a small diameter on a bunsen burner or by "Geloader Tips" (Eppendorf). Cells were microinjected under a phase-contrast inverted microscope (Olympus IMT2) at x 400 magnification. Constant injection pressure was supplied by an Eppendorf microinjector 5242 and ranged between 60 and 600 hPa depending on the permeability of individual glass capillaries and the viscosity of microinjected solutions. It should be noted that the application of a constant pressure led to the delivery of sample solution into the culture medium. However, HIV-1 replication was only observed when proviral DNA was directly injected into living cells.

For microinjection, spread cells with clearly visible nuclei were used which had grown tightly bound to the glass support and had already passed one cell division cycle indicating viability. The angle under which the injection capillary entered the nucleus (45-60°), as well as the injection time (0.1-1.0 seconds), were chosen such that the cellular and nuclear membranes were damaged as little as possible. The tip of the injection capillary was moved downward on top of the left third of the nucleus (Figure 1) to guarantee reproducible entry into the nucleus. The optimal injection volume can be estimated by the progressively changing size and contrast of the nucleus in relation to the cytoplasm.

Quantitative Measurement of HIV-1 Replication

Indirect immunofluorescence. Cells were grown on microscope slides, air-dried 24 hours after microinjection and fixed with precooled (-20°C) acetone for 20 minutes. The detection of HIV-1 specific antigens or of the *rep* gene was performed with an AIDS patient serum or a polyclonal rabbit anti-*rep*-antiserum respectively diluted with phosphate buffered saline. The second, anti-human- or anti-rabbit-directed antibodies were covalently linked with fluorescein isothiocyanate (FITC). The percentage of cells being indirect immunofluorescence-positive for HIV-1 antigens (infected cells) could be determined reproducibly in a range between 0.1% and 20%. At percentages of indirect immunofluorescence-positive cells greater than 20% a stage of HIV-1 infection is reached where cell death and cell destruction begin to appear.

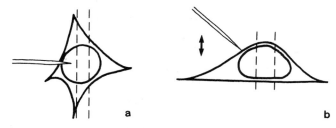

a   b

Figure 1. Schematic drawing of the movement of the injection capillary into the nucleus of a microinjected cell (double-headed arrow). The dashed lines represent auxiliary planes subdividing the cell area of capillary entry into the cell: (**a** - vertical view, **b** - horizontal view).

HIV-1 Antigen Determination

As a measure for HIV-1 replication we determined the release of HIV-1 specific antigens from co-cultured SW480/MT-4 cells four days after microinjection. The concentrations of HIV-1 antigens in dilutions of cell-free co-culture supernatants with phosphate buffered saline were measured using a commercial HIV-1 antigen ELISA (Organon-Teknika, Holland). This ELISA recognizes preferentially mature *gag* antigens and *env* proteins.

RESULTS AND DISCUSSION

Assay System

The principle of this assay (Figure 2) is based on the observation that microinjection of infectious proviral HIV-1 DNA into the nuclei of human fibroblasts and epitheloid cells leads to viral replication and the production and release of infectious HIV-1 (Boyd *et al.*, 1988; Sczakiel *et al.*, 1990). Therefore, the inhibitory effects of any potential antiviral molecule including nucleic acids, proteins and low molecular weight compounds can be tested and related to each other. In this study, eucaryotic expression plasmids or antisense RNA synthesized *in vitro* were co-microinjected together with cloned proviral HIV-1 DNA (pNL43, Adachi *et al.*, 1986) at defined amounts and molar ratios into the nuclei of 20 SW480 cells. These cells can replicate, i.e. produce HIV-1 after transfection or microinjection of proviral DNA, but cannot be infected, probably due to lack of the CD4 antigen, the main HIV-1 receptor. Virus initially produced in microinjected cells was measured after amplification with co-cultivated $CD4^+$ MT-4 cells by indirect immunofluorescence staining or by HIV-1 antigen ELISA four days after microinjection.

As shown in Figure 2b the HIV-1 antigen concentrations are proportional to the concentration of proviral HIV-1 DNA microinjected initially, thus enabling one to test inhibitory effects of co-microinjected test substances in a linear range of the assay.

The concentration of proviral DNA in this assay seems to be moderate. Assuming an injection volume in the range of 10^{-14} l per injected nucleus (Graessmann *et al.*, 1980; Sczakiel *et al.*, 1990), a pNL43 concentration of 1 ng/µl should lead to 1 molecule pNL43-DNA per cell. The absence, or a low level, of HIV-1 antigen production was measured when the pNL43 concentration was smaller than 1 ng/µl. This indicated that the injection volume was estimated to be in the correct range or that more than one pNL43 molecule per cell is necessary for productive infection.

The error range in HIV-1 replication experiments expressed as the 1 x standard deviation is in the range of 20% to 30% for 20 microinjected cells per dish. However, when the expression of an indicator gene was measured instead of HIV-1 replication, the error range expressed as the 1 x standard deviation decreased significantly to levels of equal to or smaller than 10% (see also chapter by Amet and White in this volume).

Antisense RNA-mediated Inhibition of HIV-1 Replication

Antisense nucleic acids, i.e. single-stranded nucleic acids complementary to given single-stranded target nucleic acids, can act *in vitro* and *in vivo* as specific down-regulators for gene expression and viral replication (for reviews see: Weintraub, 1990; Hélène and

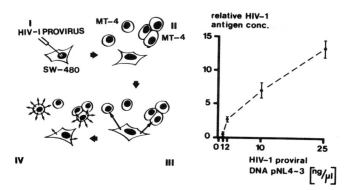

Figure 2. Principle of the microinjection assay for HIV-1 replication as described in detail by Sczakiel *et al.* (1990). **a) I.** Monolayer SW480 cells are co-microinjected with infectious proviral HIV-1 DNA and test substances. **II.** CD4$^+$ MT-4 cells which grow in suspension are co-cultivated. **III.** HIV-1 released from initially microinjected SW480 cells infect MT-4 cells. **IV.** HIV-1 replication in MT-4 cells leads to measureable amounts of cell free virus, i.e. HIV-1 antigens. **b)** Correlation between the concentration of proviral DNA and the culture concentrations of HIV-1 specific antigens as measured by ELISA four days after microinjection of 20 SW480 cells.

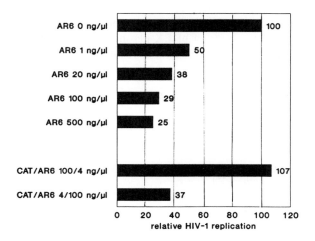

Figure 3. Antisense RNA-mediated inhibition of HIV-1 replication. Proviral DNA (pNL43, 10 ng/µl) was co-microinjected with *in vitro* synthesized HIV-1-directed antisense RNAs (AR6) or control RNAs derived from the bacterial *cat* gene at the indicated concentrations.

Toulmé, 1990). Antisense nucleic acids were also tested as inhibitors of HIV-1 replication. The bulk of this work was performed with various chemically modified or unmodified antisense oligonucleotides which were simply added to the culture medium of test cells (Zamecnik *et al.*, 1986; Matsukura *et al.*, 1987). However, there was only a very limited number of investigations of intracellular antisense RNA directed against HIV-1 (for review see: James, 1991).

We first tested the effects of *in vitro*-synthesized antisense RNA on HIV-1 replication in the microinjection assay. The local viral target region to which the 562 nucleotides long antisense construct AR6 (Rittner and Sczakiel, 1991) was directed contains coding exons for the viral regulators *tat* and *rev* as well as *vpu*, *vpr* and *env*. The resulting inhibition data showed a concentration-dependent increase in inhibition of HIV-1 replication (Figure 3). To exclude the possibility that this effect was simply due to an increase of the applied RNA concentrations, we tested mixtures of RNA containing antisense RNA (AR6) and the non-inhibitory chloramphenicol acetyl transferase (CAT) RNA with constant total RNA concentrations. Inhibition of HIV-1 replication was only seen for the mixture containing 100 ng/µl antisense RNA and 4 ng/µl CAT RNA, whereas the mixture containing 100 ng/µl CAT RNA and only 4 ng/µl antisense RNA was not inhibitory (Figure 3).

In the microinjection assay, the maximal inhibition rarely exceeds 80% (see also Figure 3). The reasons for this phenomenon which is not observed with other transfection protocols (e.g., liposome-mediated co-uptake and Ca-phosphate co-precipitation of antisense RNA/pNL43 mixtures) are not known so far. However, the microinjection assay served for reliable predictions for the antiviral effectiveness of antisense constructs used in stably transfected human T-cell lines (Sczakiel *et al.*, 1991; Sczakiel *et al.*, 1992).

One possible and promising future application of the antisense principle in the case of infection with HIV-1 is the stable intracellular expression of HIV-1-directed antisense RNA which could possibly lead to "intracellular immunization" against HIV-1 replication (Baltimore, 1988). This approach is based on effective antisense genes which were also tested transiently in the microinjection assay. The eucaryotic HCMV IE promoter/enhancer-driven expression cassette of pKEX plasmids (Rittner *et al.*, 1991a) was used to insert HIV-1 sequences in antisense orientation. A number of 10 different constructs were

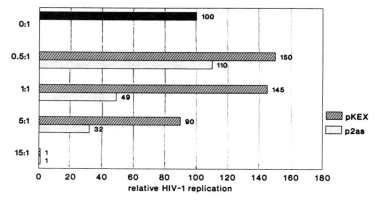

Figure 4. HIV-1 replication in the presence of increasing amounts of the antisense RNA expression plasmid p2as or the control expression plasmid pKEX (given as molar ratios to pNL43).

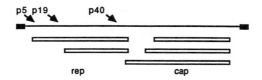

Co-microinjected plasmids	AAV-2 genes		relative HIV-1 replication [%]
	rep	cap	
pNL43	–	–	100.0 ± 28.0
pNL43 + pTAV2	+	+	3.3 ± 1.0
pNL43 + pTAV2-3	–	+	73.0 ± 16.0
pNL43 + pTAV2-6	+	–	0.9 ± 0.5
pNL43 + pTAV2-8	–	–	97.0 ± 29.0

Figure 5. AAV-2 *rep* gene-mediated inhibition of HIV-1 replication. Structure of AAV-2 wild type genome. The AAV-2 genome is represented schematically with the inverted terminal repeats serving as origins of replication (filled boxes). The open reading frames encoding the nonstructural proteins (*rep*) and three capsid proteins (*cap*) are indicated by open boxes. The two spliced forms of rep are not mentioned. Construction of the AAV-2 mutants (pTAV constructs) is explained in Heilbronn *et al.* (1990). Inhibition of HIV-1 replication is dependent on an intact *rep* gene open reading frame. For the co-microinjection of infective proviral HIV-1 DNA (pNL43) and AAV-2 mutants the following concentrations were used: AAV DNA: 100 ng/µl and HIV-1 proviral DNA pNL43: 10 ng/µl in 10 mM Tris-buffer, pH 8.0 and 1 mM EDTA. The HIV-1 replication (and 1 x standard deviation) gives the mean of four (pNL43) and eight (all pNL43/pTAV mixtures) independent co-microinjection experiments respectively.

tested in this assay and essentially two constructs (p2as and pAR6) showing significant inhibition of HIV-1 replication of up to 80% were identified (Rittner and Sczakiel, 1991). Subsequently, the antisense RNA expression plasmid p2as was tested in a concentration dependent manner which led to the observation that, beside antisense RNA-mediated inhibition of HIV-1 replication, there was also inhibition caused by other plasmid-located sequence, elements. In Figure 4 it is shown that the parental expression plasmid pKEX, which does not contain any HIV-1 sequence, led to inhibition of HIV-1 replication. An analysis of the functional sequence elements contained on pKEX showed that the inhibition could be assigned to the presence of the promoter/enhancer element (Sczakiel *et al.*, 1990).

rep Gene-mediated Inhibition of HIV-1 Replication

The microinjection assay was also used to investigate the possible interference between the adeno-associated virus type 2 (AAV-2) and HIV-1. AAV-2 is a human

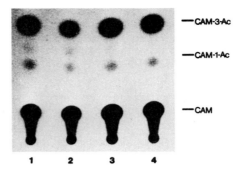

Figure 6. Autoradiography (7 days exposure time) of the expression of a HIV-1
LTR-driven *cat* gene (pHIVLTR-CAT, 50 ng/µl) 24 hours after microinjection in 50
SW480 cells (4 individual experiments).

parvovirus with a single-stranded 5 kb DNA genome consisting of three main genetic
elements:-

a) the terminal repeats which contain partially double stranded structures serving as origin
of replication,

b) the *cap* gene encoding the three structural proteins, and

c) the *rep* gene.

The latter encodes four non-structural proteins (78 kD, 68 kD, 52 kD, and 40 kD)
which are involved in AAV-2 replication and gene expression (Tratschin *et al.*, 1986;
Beaton *et al.*, 1989). AAV-2 has a broad spectrum of trans-regulating activities in different
biological systems including inhibitory effects on functions of other viruses (for review see:
Berns, 1990). Because many of the AAV-2-mediated effects can be assigned to the *rep* gene
of AAV-2 we studied the possible interference of AAV-2 and HIV-1 with particular view
on the role of the AAV-2 *rep* gene.

Co-introduction of AAV-2 DNA together with HIV-1 DNA (pNL43) in
microinjected cells (Figure 5) showed that complete cloned AAV-2 DNA inhibited
replication of HIV-1 by more than 95% (Rittner *et al.*, 1991b). Further, mutational analyses
of the AAV-2 genome showed that the AAV-2 encoded rep gene was necessary for this
effect which was also found by others in co-transfection experiments (Antoni *et al.*, 1991).
Mutants of the Rep p78 protein which were expressed from pKEX-derived expression
plasmids showed that an intact ATP binding site and portions between the start site of p78
rep and p52 *rep* were required for inhibition (Rittner *et al.*, 1992).

Expression of the *cat* Gene

The microinjection assay for quantitative measurements of HIV-1 replication has
been shown to lead to reproducible inhibition data when virus replication was monitored.
Since virus replication after transfection or microinjection of infectious nucleic acids is very
sensitive, we further investigated whether the expression of an indicator gene driven by a
weak promoter could be followed after microinjection. Thus, a HIV-1 LTR-driven CAT
expression plasmid was microinjected into the nuclei of 50 SW480 cells at a concentration

of 50 ng/µl. After two days, cells were isolated, washed, lysed and CAT-activity was measured with total protein extracts following the protocol described by Rittner *et al.* (1991a). The results shown in Figure 6 indicated that even for the HIV-1 LTR which is a very weak promoter in the absence of *Tat*, expression of the CAT gene could be detected. At the time point of protein extraction from microinjected cells, i.e. 24 hours after microinjection, a CAT activity of 8.8×10^{-8} U/cell was found which corresponds to 6600 CAT-molecules per cell.

CONCLUSION

The microinjection assay for HIV-1 replication as described here has been shown to represent a useful alternative to other transient replication assays. For example, the possible interference of AAV-2 with HIV-1 replication has been studied by co-infection and cotransfection experiments without any clear outcome, whereas the co-microinjection of cloned DNA of both viruses led to unequivocal results.

In this work, a manual microinjector was used although automatic systems for capillary microinjection have been developed and applied successfully (Ansorge and Pepperkok, 1988; Pepperkok *et al.*, 1988) and might lead to minimized error ranges. For our purposes, where relatively small numbers of cells (e.g., 20 cells per dish) are microinjected with a large number of different test solutions, glass capillaries have to be exchanged frequently and the visual system has to be refocused for each sample. Similar manipulations have to be performed in the use of an automatic microinjector. Thus, manual microinjection does not take significantly more time.

REFERENCES

Adachi, A., Gendelman, H.E., König, S., Folks, T., Willey, R., Rabson, A., and Martin, M.A. (1986) Production of acquired immunodeficiency syndrome-associated retrovirus in human and nonhuman cells transfected with an infectious molecular clone. *J. Virol.* **59**: 284-291

Ansorge, W., and Pepperkok, R. (1988) Performance of an automated system for capillary microinjection into living cells. *J. Biochem. Biophys. Methods* **16**: 283-292

Antoni, B.A., Rabson, A.B., Miller, I.L., Trempe, J.P., Chejanowsky, N., and Carter, B.J. (1991) Adenoassociated virus *rep* protein inhibits human immunodeficiency virus production in human cells. *J. Virol.* **65**: 396-404

Baltimore, D. (1988) Intracellular immunization. *Nature* **335**: 395-396

Barre-Sinoussi, F., Cherman, J.C., Rey, R., Nugeryre, M.T., Chamaret, S., Gruest, J., Dauget, C., Axler-Blin, C., Vernizet-Brun, F., Rouzioux, C., Rosenbaum, W., and Montagnier, L. (1983) Isolation of a T-lymphotropic retrovirus from a patient at risk for acquired immunodeficiency syndrome (AIDS). *Science* **220**: 868-871

Beaton, A., Palumbo, P., and Berns, K.I. (1989) Expression from the adeno-associated virus p5 and p19 promoters is negatively regulated in trans by the *Rep* protein. *J. Virol.* **63**: 4450-4454

Berns, K.I. (1990) Parvovirus replication. *Microbiol. Reviews* **54**: 316-329

Boyd, A.L., Wood, T.G., Buckley, A., Fischinger, P.J., Gilden, R.V., and Gonda, M.A. (1988) Microinjection and expression of an infectious proviral clone and subgenomic envelope construct of a human immunodeficiency virus. *AIDS Res.Hum.Retroviruses* **4**: 31-41

Gnirke, A., and Huxley, C. (1991) Transfer of the human HPRT and GART genes from yeast to mammalian cells by microinjection of YAC DNA. *Somat. Cell Mol. Genet.* **17**: 573-580

Graessmann, A., Graessmann, M., and Mueller, C. (1980) Microinjection of early SV40 DNA fragments and T antigen. *Methods in Enzymol.* **65**: 816-825

Harada, S., Koyanagi, Y., and Yamamoto, N. (1985) Infection of HTLV-III/LAV in HTLV-I-carrying cells MT2 and MT-4 and application in a plaque assay. *Science* **229**: 563-566

Heilbronn, R., Bürkle, A., Stephan, S., and zur Hausen, H. (1990) The adeno-associated virus *rep* gene suppresses herpes simplex virus-induced DNA amplification. *J. Virol.* **64**: 3012-3018

Hélène, C., and Toulmé, J.J. (1990) Specific regulation of gene expression by antisense, sense and antigene nucleic acids. *Biochim. Biophys. Acta* **1049**: 99-125

James, W. (1991) Towards gene inhibition therapy: a review of progress and prospects in the field of antiviral antisense nucleic acids and ribozymes. *Antiviral Chem. Chemotherapy* **2**: 191-214

LaMorte, V.J., Goldsmith, P.K., Spiegel, A.M., Meinkoth, J.L., and Feramisco, J.R. (1992) Inhibition of DNA synthesis in living cells by microinjection of Gi2 antibodies. *J. Biol. Chem.* **267**: 691-694

Leibovitz, A., Stinson, J.C., McCombsIII, W.B., McCoy, C.E., Mazur, K.C., and Mabry, N.D. (1976) Classification of human colorectal adenocarcinoma cell lines. *Cancer Res.* **36**: 4562-4569

Matsukura, M., Shinozuka, K., Zon, G., Mitsuya, H., Reitz, M., Cohen, J.S., and Broder, S. (1987) Phosphorothioate analogs of oligodeoxynucleotides: inhibitors of replication and cytopathic effects of human immunodeficiency virus. *Proc. Natl. Acad. Sci. USA* **84**: 7706-7710

Modalsli, K.R., Bukholm, G., Mikalsen, S.-O., and Degré, M. (1991) Microinjected coxsackie B1 virus does not replicate in HEp-2 cells. *Virology* **185**: 888-890

Pepperkok, R., Zanetti, M., King, R., Delia, D., Ansorge, W., Philipson, L., and Schneider, C. (1988) Automatic microinjection system facilitates detection of growth inhibitory mRNA. *Proc. Natl. Acad. Sci. USA,* **85**: 6748-6752

Popovic, M., Sarngadharan, M.G., Read, E., and Gallo, R.C. (1984) Detection, isolation and continous production of cytopathic retroviruses (HTLV-III) from patients with AIDS and pre-AIDS. *Science* **224**: 497-500

Riabowol, K.T., Vosatka, R.J., Ziff, E.B., Lamb, N.J., and Feramisco, J.R. (1988) Microinjection of fos-specific antibodies blocks DNA synthesis in fibroblast cells. *Mol. Cell. Biol.* **8**: 1670-1676

Rittner, K., and Sczakiel, G. (1991) Identification and analysis of antisense RNA target regions of the human immunodeficiency virus type 1. *Nucl. Acids Res.* **19**: 1421-1426

Rittner, K., Stöppler, H., Pawlita, M., and Sczakiel, G. (1991a) Versatile eucaryotic vectors for strong and constitutive transient and stable gene expression. *Methods Mol. Cell. Biol.* **2**: 176-181

Rittner, K., Heilbronn, R., Kleinschmidt, J.A., Oelze, I., and Sczakiel, G. (1991b) Replication of the immunodeficiency virus type 1 is inhibited by the adeno-associated virus type 2 *rep* gene. *Biochem. Soc. Trans.* **19**: 438

Rittner, K., Heilbronn, R., Kleinschmidt, J.A., and Sczakiel, G. (1992) Adeno-associated virus type 2-mediated inhibition of human immunodeficiency virus type 1 replication: involvement of $p78^{rep}/p68^{rep}$ and the HIV-1 LTR. J. Gen. Virol. (in press)

Sczakiel, G., Pawlita, M., and Kleinheinz, A. (1990) Specific inhibition of human immunodeficiency virus type-1 replication by RNA transcribed in sense and antisense orientation from the 5'-leader/gag region. *Biochem. Biophys. Res. Comm.* **169**: 643-651

Sczakiel, G., Rittner, K., and Pawlita, M. (1991) Human immunodeficiency virus type 1 replication is reduced by intracellular antisense RNA expression, *in:* "Prospects for antisense nucleic acid therapy of cancer and AIDS", E.Wickstrom, ed., Wiley-Liss, New York, 179-193

Sczakiel, G., Oppenländer, M., Rittner, K., and Pawlita, M. (1992) Tat- and Rev-directed antisense RNA expression inhibits and abolishes replication of the human immunodeficiency virus type 1: a temporal analysis. *J. Virol.* (in press)

Smith, M.R., Ryu, S.-H., Suh, P.-G., Rhee, S.-G., and Kung, H.-F. (1989) S-phase induction and transformation of quiescent NIH 3T3 cells by microinjection of phospholipase C. *Proc. Natl. Acad. Sci. USA* **86**: 3659-3663

Tratschin, J.D., Tal, J., and Carter, B.J. (1986) Negative and positive regulation in trans of gene expression from adeno-associated virus vectors in mammalian cells by a viral rep gene product. *Mol. Cell. Biol.* **6**: 2884-2894

Wawra, E. (1988) Microinjection of deoxynucleotides into mouse cells. *J. Biol. Chem.* **263**: 9908-9912

Weintraub, H.B. (1990) Antisense RNA and DNA. *Sci. Am.* **262**: 34-40

Zamecnik, P.C., Goodchild, J., Taguchi, Y., and Sarin, P.S. (1986) Inhibition of replication and expression of human T-cell lymphotropic virus type III in cultured cells by exogenous synthetic oligonucleotides complementary to viral RNA. *Proc. Natl. Acad. Sci. USA* **83**: 4143-4146

HOW TO MAKE GLASS MICROTOOLS FOR THE

INJECTION OF ISOLATED PLANT SPERM CELLS

INTO EMBRYO SAC CELLS USING A

MICROFORGE

C.J. Keijzer

Department of Plant Cytology and Morphology, Agricultural University Wageningen, Arboretumlaan 4, 6703 BD Wageningen, The Netherlands

INTRODUCTION

Preceding fertilization in higher plants, the pollen tube penetrates a synergid of the embryo sac in order to shed its contents, including the two sperm cells. In order to analyse additional functions of the pollen tube cytoplasm in the synergid (apart from sperm cell transfer), we directly injected isolated sperm cells into synergids of *Torenia fournieri*, using micromanipulation (Keijzer *et al.*, 1988a; Keijzer, 1992). The rate of successful sperm cell transfers in these studies was, however, low (Keijzer, 1992) but serve to illustrate the potential usefulness of the technique to be described below.

MATERIALS AND METHODS

Although the different steps of this difficult artificial fertilization procedure have been described (Keijzer, 1992), the detailed description of the developed micromanipulation and microforge techniques have not yet been presented. They will be presented here since they may be useful in other micromanipulation projects.

Biotechnology Applications of Microinjection, Microscopic Imaging, and Fluorescence, Edited by P.H. Bach *et al.*, Plenum Press, New York, 1993

11

Sperm Cell Isolation

A detailed description of the sperm cell isolation procedure has been published (Keijzer *et al.*, 1988a; Keijzer, 1992). In brief, decreasing the osmotic pressure by diluting a liquid pollen germination medium makes the pollen tubes burst, thus releasing the sperm cells. Increasing the osmotic pressure again, by adding a high saccharose concentration, prevents the sperm cells from bursting (see the reviews of Keijzer *et al.*, 1988b and Theunis *et al.*, 1991). The sperm cells are then sucked into a special micropipette, the fabrication of which is described in the next section, in order to inject them into an embryo sac cell.

Making an Injection Pipette

To penetrate the tough wall of the embryo sac, a very sharp pipette tip is needed. As a consequence, this tip cannot be the site of the aperture to transfer the 3 μm-wide sperm cells. Therefore, we made microinjection pipettes with a sharp closed tip (for penetration) and a lateral aperture for the transfer as shown (Figure 1).

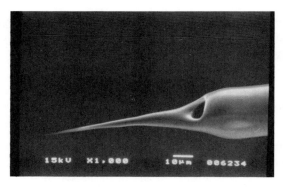

Figure 1. SEM-photograph of a micropipette used for the injection of sperm cells into embryo sac cells.

De Fonbrune (1949) developed procedures to make such pipettes, using his microforge. However, his methods lead to pipettes with a minimum distance between the tip and the aperture of about 200 μm, i.e. too large for our purposes. Using an advanced microforge, derived from his original design, we developed a procedure to make this distance shorter (Figures 1 and 2).

Using a Leitz horizontal pipette-puller, a micropipette with an aperture diameter of about 1 μm is made out of a Pasteur-pipette. This pipette is placed in a vertical position with the tip upward in the pipette clamp of the microforge. Now a 1.2 bar flow of carbon dioxide is pressed through the pipette and the outside of the tip is cooled with a gentle stream of air at room temperature (Figure 2a). Subsequently, the platinum heating filament is melted to one side of the rim of the pipette aperture (Figure 2b) and rapidly retracted (Figure 2c), thus creating a massive penetration tip (Figure 2d). Since tips drawn in this way are too thin to

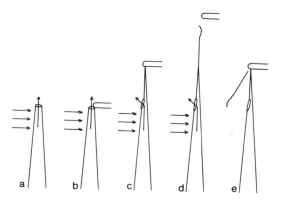

Figure 2. Schematic drawing of the fabrication of the pipette of Figure 1 using a microforge. **a.** Switching on an air-flow from the left and a carbon dioxide-flow of 1.2 bar through the pipette. **b.** The heated filament is melted to one side of the rim of the pipette. **c.** The filament draws the pipette tip into a thin thread, the entire pipette aperture melts and is drawn aside and enlarges. **d.** The movement of the previous step is extended, drawing the tip into an ultra-thin thread, which finally disconnects from the filament and bends due to the heat of the latter. **d.** Using the cold filament, the tip is broken at a site where it is rigid enough to penetrate the cell wall of the female target cells.

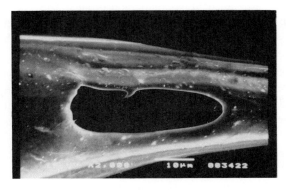

Figure 3. A detailed SEM-photograph of a lateral aperture of an injection pipette. Note the tiny glass hook, which makes this pipette useless.

be rigid (diameter of the extreme tip is about zero), they are broken a few μm behind the tip using the cold filament (Figure 2e). During these melting and drawing steps the entire area around the original aperture is melted as well and drawn to a more or less lateral position behind the newly formed tip (Figures 2c, d). Due to the internal carbon dioxide flow, it remains open during this operation. The result is shown in Figure 1.

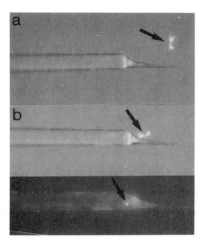

Figure 4. An isolated sperm cell pair (arrow), stained with the fluorescent DNA-probe DAPI, is sucked into the pipette of Figure 1.

Figure 3 shows a detailed SEM-photograph of such a lateral aperture. Before use each pipette must be checked in this microscope. Apart from exactly determining the sizes, tiny fabrication faults can be detected, which cannot be observed with the light microscope, but are big enough to destroy the sperm cells.

An additional advantage of such a relatively large aperture is the low capillary suction of the pipette. For our apertures with diameters of 5-10 μm, the sucking force of a human mouth is sufficient for picking up sperm cells from their isolation medium (Figure 4). This allows for very careful control of the sucking and blowing work.

Making Microforceps Beaks

The ovule, which is isolated mechanically from the placenta using preparation needles and a stereo microscope (Keijzer *et al*, 1988a; Keijzer, 1992), is kept in position using microforceps. Depending upon the goal of the microinjection work, two types of forceps beaks can be made with the microforge. First the procedure of Figure 5 can be followed, leading to a microfork. Therefore, a 1 μm tip micropipette is made using the pipette puller. It is clamped in a vertical position in the microforge and cooled at its outside with a flow of air at room temperature. Now tiny lateral spicules are made one by one by shortly melting the heated filament to a site shortly behind the tip of the pipette (Figure 5a) and rapidly retracting it by which the melting glass is pulled aside and finally detaches from the filament to form a fine spicule (Figure 5b). If this is repeated a few times, a fork is made (Figures 5c-d, 6). When two such forks are used as the beak of a microtweezer, they guarantee a well-fixed ovule, the epidermis of which may be slightly damaged due to penetration by some of the spicules.

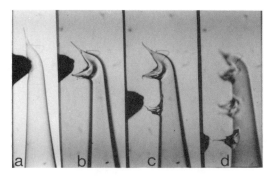

Figure 5. Photographs from the microforge showing the fabrication of a fork-shaped beak half to be used in microforceps. While air is blown from the right, the heated filament draws spikelets from the (other) lateral side of the micropipette.

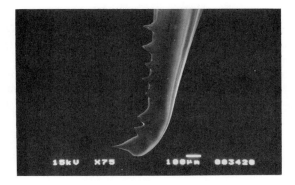

Figure 6. SEM-photograph of a microforceps beak half made according to the procedure in Figure 5.

A second possibility is to use two cup-shaped beak halves, that exactly fit on two opposite sides of the ovule (Figure 7). These beaks can be made according to the procedure of De Fonbrune (1949), which is shown in Figure 8. Therefore, a 1 µm tip micropipette is placed in a vertical position in the microforge and a stream of air at room temperature is blown from aside. Now the tip of the pipette is locked by approaching it with the heated filament of the microforge (Figure 8a). Subsequently, carbon dioxide at 1.2 bar is pressed into the pipette. When the closed tip is approached by a heated filament, the former is blown up into a small sphere (Figure 8b). Now the carbon dioxide pressure is stopped and the slightly heated filament is melted to a lateral side of the sphere (Figure 8c). If the heating is switched off, the filament retracts due to its own cooling/shrinkage, thus breaking the sphere into two halves (Figure 8d). The edges of the sphere half which is still connected to the pipette are made smooth by approaching them with the warm filament (Figure 8e). Now the beak is ready for use (Figure 7).

The advantage of using microforceps to keep the ovule in position, when compared with an agarose layer, is the possibility to transfer such a single ovule afterward into any desired medium or fixative.

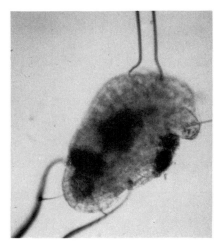

Figure 7. For microinjection of sperm cells, the isolated ovules of *Torenia fournieri* can be very well fixed into position in microforceps with cup-shaped beaks.

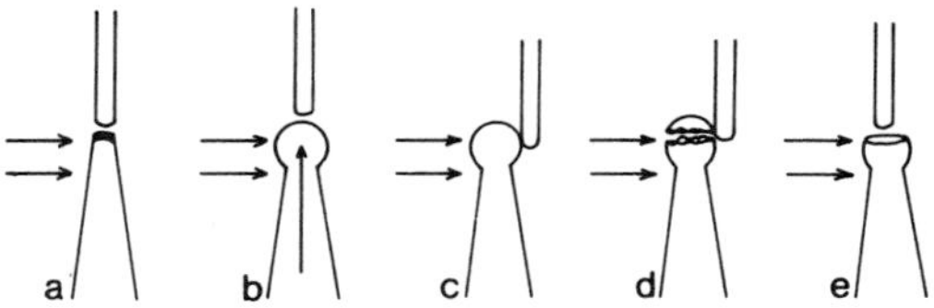

Figure 8. Fabrication of a cup-shaped microforceps beak-half using a microforge. **a.** A flow of air is blown from the left, and the tip of a 1 µm micropipette is closed by approaching it with the warm filament. **b.** Carbon dioxide gas at 1.2 bar is pressed into the pipette, blowing the melted tip of the pipette into a sphere. **c.** The carbon pressure is stopped and the slightly heated filament is melted to a lateral side of the sphere. **d.** When the heating is switched off, the filament shrinks slightly, breaking the sphere at the attachment site. **e.** The sharp edges of the remaining sphere half are made blunt by approaching them with a warm filament.

The advantage of using microforceps over a (sucking) holder pipette, which is generally used for protoplast or pollen grain injections, is the absence of streaming in the medium. This may cause debris from the ovule isolation procedure to be sucked towards the ovule during the careful injection work. This might cause fatal vibration.

The Transfer of the Sperm Cells into the Embryo Sac

The injection of the embryo sac is the most difficult step of the procedure (Figure 7; Keijzer, 1992). Penetrating its wall frequently led to explosion of the (plasmolysed) cell, which was the main reason for the low number of successful sperm cell transfers in our experiments (Keijzer, 1992). Despite the sharp, aperture-less tip of the injection pipette, the tough embryo sac wall generally did not immediately let it pass through. Generally the wall was first bent strongly inward, thus increasing the pressure inside the cells, until their resistance was strong enough for the pipette to (suddenly) penetrate them. Apart from penetrating the ovule too deeply, this overruled the effect of the plasmolysis and the cytoplasm was expelled from the embryo sac.

CONCLUSIONS

Nowadays microforges are generally used to make rather simple tools like holder pipettes to stabilize cells and organs during micromanipulation work. The many possibilities outlined in the "bible of microforge work", written in French by De Fonbrune (1949), are seldomly used in biological research. The main reason is that these possibilities are not well known among most researchers who use micromanipulation. Secondly, working with these specialized microtools takes lots of time and patience. On the other hand, the fabrication of these specialized microtools in the microforge appears to be rather simple. In spite of the thorough studies of De Fonbrune, it appears to be possible to make the tools even smaller than he described. Such refined tools can be easily made, but they probably approach the limits of what is possible with glass microtools.

REFERENCES

De Fonbrune, P. (1949) Technique de Micromanipulation, Masson, Paris

Keijzer, C.J., Reinders, M.C., and Leferink-ten Klooster, H.B. (1988a) A micromanipulation method for artificial fertilization in *Torenia*, *in:* "Sexual Reproduction in Higher Plants", M. Cresti, P. Gori and E. Pacini, eds., Springer, Berlin, 119-124

Keijzer, C.J., Wilms, H.J., and Mogensen, H.L. (1988b) Sperm cell research, the current status and applications for plant breeding, *in:* "Plant Sperm Cells as Tools for Biotechnology", H.J. Wilms and C.J. Keijzer, eds., Pudoc, Wageningen, 3-8

Keijzer, C.J. (1992) The isolation of sperm cells, their microinjection into the egg apparatus and methods for structural analysis of the injected cells, *in:* "Promotion of Sexual Plant Reproduction", M. Cresti and A. Tiezzi, eds., Springer, Berlin

Theunis, C.H., Pierson, E.S., and Cresti, M. (1991) Isolation of male and female gametes in higher plants. *Sex. Plant Reprod.* **4**: 145-154

APPLICATION OF THE FIREFLY LUCIFERASE

REPORTER GENE TO MICROINJECTION

EXPERIMENTS IN *XENOPUS* OOCYTES

Michael R.H.White[1,2], Martin Braddock[1], Elaine D. Byles[1], Lorene Amet[1], Alan J. Kingsman[1,3], and Susan M. Kingsman[1]

[1]Virus Molecular Biology Group, Department of Biochemistry, South Parks Road, Oxford, OX1 3QU, UK

[2]Amersham International plc, White Lion Road, Little Chalfont, Bucks, HP7 9LL, UK

[3]British Bio-technology Ltd, Watlington Road, Oxford, OX4 5LY, UK

INTRODUCTION

The human immunodeficiency virus (HIV-1) is the aetiological agent of the acquired immune-deficiency syndrome (AIDS) (Barre-Sinoussi *et al.*, 1983; Gallo *et al.*, 1984). HIV-1 has the typical genetic organisation of retroviruses, with three major genes gag, pol and env. In addition, HIV-1 has additional short open reading frames that encode various regulatory proteins making it one of the most complex retroviruses that has been described (Varmus *et al.*, 1984). The regulation of gene expression in HIV-1 has been shown to be critically dependent on the virally-encoded TAT protein (Dayton *et al.*, 1986).

The TAT protein is a potent stimulator of viral gene expression by an entirely novel mechanism which has not yet been fully elucidated. TAT activation gives rise to increases in the level of viral gene expression from the HIV-1 long terminal repeat and, therefore, increases the rate of its own synthesis and the synthesis of all viral proteins. The TAT protein has been shown to bind to a cis-acting element called TAR, which is present in the 5′ untranslated regions of all HIV mRNAs (Rosen *et al.*, 1985). TAR RNA forms a stable bulge stem-loop structure (Muesing *et al.*, 1987; Roy *et al.*, 1990) and *in vitro* studies have

Biotechnology Applications of Microinjection, Microscopic Imaging, and
Fluorescence, Edited by P.H. Bach *et al.*, Plenum Press, New York, 1993

shown that the bulge is necessary for TAT binding (Roy *et al.*, 1990). TAT acts predominantly at the level of transcription in mammalian cells (Rice and Matthews, 1988; Jakobovits *et al.*, 1988). However, some studies have suggested that TAT may function post-transcriptionally in mammalian cells to increase the translational capacity of TAR RNAs (Muesing et al., 1987; Rosen et al., 1986).

In order to investigate the possible post-transcriptional mechanism of trans-activation by the TAT protein, we have used the *Xenopus laevis* oocyte model system (Braddock *et al.*, 1989, 1990, 1991). The ease and accuracy of microinjection of biological material into these very large cells has allowed us to separate the roles of different cell compartments in the regulation of reporter gene expression from *in vitro* transcribed RNA molecules. For these studies the chloramphenicol acetyl transferase (CAT) reporter gene was used, since for many years it has been the most widely used and trusted reporter of gene expression (Gorman *et al.*, 1982). The CAT coding sequence was placed downstream from the SP6 bacteriophage polymerase promoter sequence and RNAs transcribed *in vitro*, which either contained the short CAT 5′ untranslated leader sequence, or additionally a portion of the HIV-1 leader sequence which contains the TAR element. Injection of these RNA molecules into *Xenopus* oocytes showed that in the absence of the TAR element there was a high level of CAT expression when the RNA was injected into the cytoplasm. This expression was abolished when the RNA was injected into the nucleus. This nuclear inhibition of CAT expression could not be overcome by co-injection of TAT protein. Nuclear or cytoplasmic injection of TAR+ RNA into the oocyte gave no detectable CAT expression. This indicated that the TAR element gave rise to a cytoplasmic block on translation of the CAT RNA, which was not relieved by co-injection of TAT protein. However, nuclear co-injection of TAR-containing RNA, together with TAT protein, gave rise to a high level of CAT expression. These results indicated that in the oocyte system, TAT activates gene expression by a post-transcriptional mechanism. This phenomenon appears to be mediated by a TAT-dependent nucleus specific chemical modification of RNA which somehow facilities translation (Braddock *et al.,* 1991).

These experiments depended on the injection of quite high concentrations (typically 3 ng) of RNA per oocyte. Under these conditions it was still not possible to detect any CAT reporter gene expression, either when CAT RNA was injected into the nucleus, or when TAR-CAT RNA was injected into either cellular compartment in the absence of TAT. Therefore, we wanted to investigate the use of an alternative reporter gene for use in these type of experiments. The establishment of a second reproducible reporter gene assay would also permit the design of dual reporter assay experiments.

The firefly luciferase reporter gene has over the last few years been increasingly applied to gene expression studies in mammalian cells (De Wet *et al.*, 1987), plants (Ow et al., 1986), and bacteria (Palomares *et al.*, 1989). In mammalian cells this reporter gene has been shown to have increased sensitivity over CAT of 100-1000 fold (Gould and Subramani, 1988). We therefore investigate the use of this reporter gene in microinjection experiments in *Xenopus* oocytes. We show here that in *Xenopus* oocytes the luciferase reporter gene gives increased sensitivity over CAT and shows the same post-transcriptional control of HIV gene expression which was originally observed with the CAT reporter gene. This allows the injection of lower levels of RNA, and the detection of the previously uncharacterised basal levels of expression. We have also observed some significant differences between the kinetics of luciferase and CAT expression following RNA injection into oocytes and some implications of these results for reporter gene assay experiments are discussed.

MATERIALS AND METHODS

Construction of Vectors for *in vitro* Synthesis of RNA

SP6-Luc RNA: A 2.6 Kb H*ind* III - B*am* HI fragment encoding the luciferase gene, SV40 small t intron and SV40 polyadenylation sequence was cloned into pSP64 (Pharmacia). The resulting vector was cleaved with H*pa* I and B*am* HI and the resulting linear vector religated with a blunt end - B*am* HI linker which contained a S*ma* I site and a run of 30 A residues. The resulting vector may be cleaved with B*am*HI so that transcription with SP6 polymerase results in a 2.6 Kb RNA molecule which contains a synthetic 30 base poly(A) tail. This vector, pMW56, is functionally equivalent to the SP6-CAT RNA vector pPE151 (Braddock *et al.*, 1989).

SP6-TAR-Luc RNA: A 2.6 Kb B*gl* II - B*am* HI fragment containing the HIV-1 TAR sequence, luciferase gene, SV40 small t intron and SV40 polyadenylation signal was cloned into the B*gl* II-B*am* HI sites of the SP6 TAR-CAT vector pPE38 (Braddock *et al.*, 1989). The resulting vector was cleaved as above with H*pa* I and B*am* HI and the polyA linker inserted into the vector. This vector, PMW55, is functionally equivalent to the TAR-CAT RNA vector pPE38 (Braddock *et al.*, 1989).

SP6-CAT RNA: Previously Constructed Vector pPE151 (Braddock *et al.*, 1989).

In vitro Transcription Reactions

Each of the above vectors were linearised with B*am* HI and the linear template purified by two phenol/chloroform and two chloroform extractions, followed by ethanol precipitation. The RNAs were produced by *in vitro* transcription with SP6 polymerase and capped as described previously (Braddock *et al.*, 1989).

Preparation and Microinjection of *Xenopus laevis* Oocytes

Stage 6 *Xenopus* oocytes were prepared as described (Braddock *et al.*, 1989). RNA (3 ng unless otherwise stated) was injected into the nucleus or cytoplasm. Titration shows that this input of RNA yields 20% of the maximum CAT activity. Where appropriate, 0.2 pg TAT protein was provided by co-injection with the RNA. This represents a 100-fold excess over the minimum amount of TAT to give maximum activation (Braddock *et al.*, 1989). TAT was produced in yeast as described (Braddock *et al.*, 1989).

Cell Lysate and Oocyte Extract Preparation

Cell lysates from various mammalian cell lines were prepared from lysis of 10^6 cells in 1 ml luciferase lysis buffer (1% v/v Triton-X100, 25 mM Tris-phosphate pH 7.75, 15% v/v glycerol, 8 mM $MgCl_2$, 0.1 mM EDTA, 1 mM DTT, 1% w/v BSA). Debris was removed from the lysate by micro-centrifugation for 2 minutes and the supernatant was stored.

Oocytes for luciferase assays were manually lysed in a modification of the above lysis solution, which lacked the Triton-X100 and BSA components (10-30 oocytes per 200 μl). Aliquots of these extracts were centrifuged for 15 minutes and the supernatant assayed for protein concentration. Other aliquots were diluted into complete luciferase lysis buffer (plus Triton-X100 and BSA) for luciferase assays. For the enzyme standard curve (Figure 1), the lysates were mixed with the appropriate concentration of purified luciferase enzyme (Boehringer) and incubated on ice for at least 15 minutes before assay.

Oocytes for CAT assays were manually lysed in 0.1 M Tris.HCl, pH 7.8. The cell debris was removed by micro-centrifugation twice for 15 minutes and the supernatant assayed for protein concentration and CAT activity.

Luciferase and CAT Assays

Aliquots of the lysates were then diluted (as appropriate) into 250 µl of lysis buffer, placed in a luminometer cuvette with 4 µl of 40 mM ATP (in 25 mM Tris-phosphate, pH 7.75) and assayed by injection of 100 µl of 2 mM luciferin (Boehringer, in 25 mM Tris-phosphate, pH 7.75) in a Lumac M2010 Biocounter in rate mode. The peak luminescent reading (light units per second) was recorded. CAT assays were performed as described (Gorman, 1982). Reactions were incubated for 1-2 hours, with typically 10 µg protein and the thin layer chromatography plates analyzed using a Phosphorimager (Molecular Dynamics).

Primer Extension Analysis of Injected and Rescued RNA

For experiments where oocytes were processed for both enzyme and RNA analysis, half of the oocytes (typically 15) were processed for each assay. RNA was rescued from oocytes using a previously described method (Braddock *et al.*, 1991). Primer extension analysis used a ^{32}P end-labelled anti-sense primer homologous to bases 9 to 33 of the luciferase coding region (De Wet *et al.*, 1987) and was performed as described previously (Braddock *et al.*, 1990).

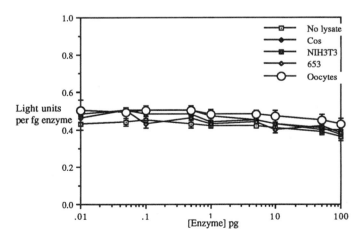

Figure 1. Reproducibility of the luciferase assay in lysate from various mammalian cell lines and from *Xenopus* oocytes. Cos = 10^6 Cos-7 (African green monkey kidney) cells per ml. NIH3T3 = 10^6 NIH3T3 (murine fibroblast) cells per ml. 653 = 10^6 X63-Ag8.653 (murine myeloma) cells per ml. Oocyte = *Xenopus* oocyte lysate adjusted to 100 µg per ml.

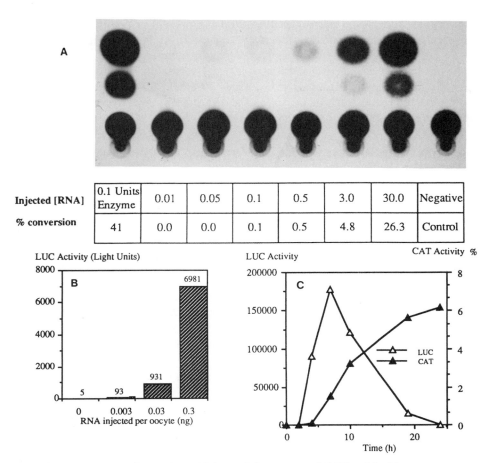

Injected [RNA]	0.1 Units Enzyme	0.01	0.05	0.1	0.5	3.0	30.0	Negative
% conversion	41	0.0	0.0	0.1	0.5	4.8	26.3	Control

Figure 2. Comparison of sensitivity and timecourse of CAT and luciferase assays following injection of RNA into *Xenopus* oocytes. **A)** CAT RNA titration (0.01-30 ng per oocyte) with RNA derived from pPE151 (Braddock *et al.*, 1989). The photograph shows a TLC plate of assays of the lysate from approximately 10 oocytes. Values below show the % conversion of [14C]chloramphenicol per oocyte 16-18 hours post-injection. **B)** Luciferase RNA titration (0.003-0.3 ng per oocyte) with RNA derived from pMW56. Luciferase activity is shown as light units per oocyte 16-18 hours post-injection. **C)** Time-course of CAT and luciferase activities following injection of 3 ng of each reporter gene RNA per oocyte. Activities are shown as light units and % conversion of [14C]chloramphenicol per oocyte.

RESULTS

Establishment of the Dynamic Range, Linearity and Reproducibility of the Luciferase Assay

In order to show that the conditions for the luciferase assay were unaffected by various cell extracts, purified luciferase enzyme was added to the cell extracts (prepared at

10^6 cells/ml) for mammalian cells and at 100 µg/ml of protein from *Xenopus laevis* oocytes. The luciferase containing extracts were then assayed in an injection luminometer for luciferase activity. Figure 1 shows a graph of the concentration of enzyme added to the different mammalian cell or oocyte extracts (X-axis) versus the resulting light emission expressed as light units per fg of enzyme assayed (Y-axis). The linearity of the assay is shown by the horizontal slope of the lines. The addition of oocyte lysate clearly had no significant effect on the assay.

Comparison of the Sensitivity of the Luciferase and CAT Assays Following Micro-injection of RNA into *Xenopus* Oocytes

In order to compare the sensitivity of the CAT and luciferase assays in *Xenopus* oocytes, we constructed vectors containing the reporter genes placed directly downstream from the SP6 bacteriophage polymerase promoter [pPE151 for CAT (Braddock *et al.*, 1989) and pMW56 for luciferase]. *In vitro* transcribed RNA was synthesised using SP6 polymerase and DNA templates linearised downstream of the reporter genes with B*am* HI. The resulting RNA was injected into pools of oocytes at varying concentrations. Figure 2a and b show the results of these RNA titrations. In each case the luciferase and CAT assays were performed 16-18 hours post-injection. These results clearly show the increased sensitivity of the luciferase assay over CAT. At 16-18 hours post-injection we could detect a signal from 100 to 1000-fold less luciferase RNA than with CAT RNA.

Comparison of the Timecourse of Luciferase and CAT Expression Following Micro-injection of RNA into *Xenopus* Oocytes

It has previously been shown that the kinetics of luciferase expression varied markedly from that of CAT following DNA transfection into mammalian cells (Maxwell and Maxwell, 1988). These kinetics were also dependent on the method of transfection. We therefore investigated the timecourse of CAT and luciferase expression post-injection of RNA into *Xenopus* oocytes. These results (Figure 2c) clearly show a marked difference, with significant luciferase expression appearing 4 hours post-injection, peaking at 7-8 hours post-injection and then decaying rapidly until 24-30 hours when the luciferase activity was undetectable. CAT activity, however, increased steadily over the 24 hours post-injection and then reached a steady plateau.

Analysis of the Post-transcriptional Effect of the TAT Trans-activator Protein on Expression of the Luciferase Reporter Enzyme Following Micro-injection of RNA into *Xenopus* Oocytes

In order to analyze the control of reporter gene expression in TAR containing RNA molecules which utilised the luciferase reporter gene instead of CAT, we constructed an SP6 vector that encoded the TAR sequence directly upstream from the luciferase gene (pMW55). Capped RNA was made from this TAR containing vector and from the TAR-luciferase vector pMW56. These RNAs were injected into the nucleus or cytoplasm of pools of *Xenopus* oocytes (3 ng RNA per oocyte). Table 1 shows examples of typical results from assays performed 16-18 hours post-injection. Injection of SP6-Luc RNA (from pMW56) into cytoplasm of oocytes (Injection A) gave rise to expression of high levels of luciferase expression as described above. When SP6-TAR-Luc RNA (from pMW55) was

Table 1. Injection of SP6-LUC and SP6-TAR-LUC RNA into the cytoplasm and nucleus of *Xenopus oocytes*.

Injection[1]	Light Signal per oocyte	Molecules of luciferase per molecule RNA
A.	1.1×10^5	1.04
B.	4×10^2	3.8×10^{-3}
C.	3.2×10^4	0.3
D.	5	-

[1]**Injection A.** 3 ng of SP6-Luc RNA injected into the cytoplasm - TAT Protein.
Injection B. 3 ng of SP6-TAR-Luc RNA injected into the cytoplasm - TAT protein.
Injection C. 3 ng of SP6-TAR-Luc RNA injected into the nucleus + TAT protein.
Injection D. No RNA.

injected into the cytoplasm the luciferase levels showed a 99.7% inhibition of luciferase activity due to the presence of the TAR element (Injection B). This inhibition of luciferase activity was unaffected by co-injection of TAT protein. These results agree with those described previously with CAT (Braddock *et al.*, 1989, 1990, 1991). The increased sensitivity of the luciferase assay allowed a low, but significant, level of luciferase activity to be detected when the SP6-TAR-Luc RNA was injected into the oocyte cytoplasm. This activity had not previously been detected with the CAT reporter gene.

When SP6-Luc RNA or SP6-TAR-Luc RNAs were injected into the oocyte nucleus, similar low levels of luciferase were observed (data not shown). This agrees with the previous CAT results which indicated a nuclear block on translation of these RNAs (Braddock *et al.*, 1989, 1990, 1991). However, when TAT protein was co-injected into the nucleus together with SP6-TAR-Luc RNA, significant activation of luciferase expression was observed (Injection C). This activation did not occur in the absence of the TAR sequence (not shown).

Primer extension analysis of the injected SP6-Luc and SP6-TAR-Luc RNAs (Figure 3) showed that approximately equal concentrations of RNA had been injected. RNA rescued from the oocytes at the same time as the luciferase assays (16-18 hours post-injection) gave rise to ^{32}P labelled products of the primer extension reactions that were of the expected sizes (Figure 3). This indicated that the 5′ terminus of the RNAs remained intact at this time interval post-injection.

DISCUSSION

The above results have shown that the firefly luciferase gene is a useful reporter gene for microinjection experiments in *Xenopus* oocytes. These experiments have confirmed the post-transcriptional activation of gene expression in oocytes by the HIV-1 TAT protein which was previously observed with the CAT reporter gene (Braddock *et al.*, 1989, 1990, 1991). The luciferase assay benefits from high sensitivity, which is significantly greater than for the CAT assay. This allows lower levels of reporter gene expression to be accurately quantified over a wide dynamic range. A particular advantage of this extra sensitivity is that it allows lower and more physiological concentrations of nucleic acids to be injected into the oocyte.

A significant difference in the timecourse of CAT and luciferase activity following microinjection of RNA was observed. The timecourse of luciferase activity was shown to be transient, reaching a peak after 7-8 hours and then falling off rapidly. The CAT activity however rose more slowly to a plateau after 24 hours which remained constant for at least 36 hours. In order to investigate this phenomenon further we studied the timecourse of the physical stability of the injected RNA by primer extension analysis (not shown). These

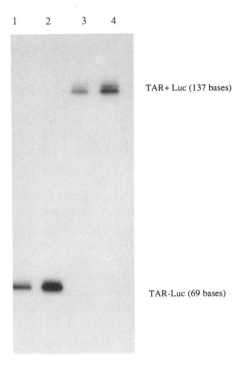

Figure 3. Primer extension analysis of SP6-Luc and SP6-TAR-Luc RNAs before injection and after rescue of the RNA from *Xenopus* oocytes.

1. TAR-RNA (pMW56 derived)	Rescued RNA from cytoplasmic injection
2. TAR-RNA	Injected RNA
3. TAR+RNA (pMW55 derived)	Rescued RNA from cytoplasmic injection
4. TAR+RNA	Injected RNA

primer extension results indicated that even 30 hours post-injection, when the luciferase activity was undetectable, the injected RNA was still physically intact.

The observation of the different time-courses of expression from RNA molecules encoding the two reporter genes is very striking. These results agree with differences in the timecourse of reporter expression which have been seen previously in mammalian cells following DNA transfection (Maxwell and Maxwell, 1988). Our primer extension results suggest that the luciferase RNAs are physically stable. Therefore the reduced time-course of luciferase expression could be due to either reduced or functional stability of the RNA or to reduced protein stability, or a combination of these factors. Other researchers have previously suggested that in mammalian cells the luciferase protein is unstable (Thompson *et al.*, 1991) and have proposed the use of luciferin analogues which protect the enzyme from proteolysis. It has also been observed from RNA transfection experiments into plant and mammalian cells that the luciferase RNA may have a relatively short half life, which is significantly shorter than the protein half-life (Gallie, 1991). The same researcher also observed that the physical half-life of the RNA was apparently longer than its functional half-life. This could be explained by the inability of the RNA assays to detect removal of one or a few bases required for function of the RNA. Further experiments to establish the stability of the luciferase protein and RNAs in *Xenopus* oocytes will be required to separate these alternative explanations for the present results.

The results described above suggest that luciferase should also be useful in experiments involving microinjection of DNA and RNA reporter constructs into mammalian cells. Luciferase has previously been used by a number of researchers as a reporter for RNA transfection into mammalian cells (Gallie, 1991; Malone *et al.*, 1989). The high sensitivity of the assay should allow the analysis of reporter gene expression in small populations of microinjected cells. A further advantage of luciferase expression in microinjected mammalian cells should be the potential for the non-destructive imaging of the luciferase activity in single mammalian cells using quantitative high sensitivity imaging cameras (White *et al.*, 1990, 1991). These techniques should allow the cell to be used as a test-tube for the analysis of gene expression.

ACKNOWLEDGEMENTS

Martin Braddock is a Royal Society Research Fellow. We would like to thank Amersham International and other members of the virus molecular biology group for support.

REFERENCES

Barre-Sinoussi, F., Chermann, J.C., Rey, R., Nugeyre, M.T., Chamaret, S., Gruest, J., Dauguet, C., Axler-Blin, C., Vezinet-Brun, F., Rouzioux, C., Rosenbaum, F., and Montagnier, L. (1983) Isolation of a T-lymphotropic retrovirus from a patient at risk for Acquired Immune Deficiency Syndrome (AIDS). *Science* **220**: 868-871

Braddock, M., Chambers, A., Wilson, W., Esnouf, M.P., Adams, S.E., Kingsman, A.J., and Kingsman, S.M. (1989) HIV-1 TAT "activates" presynthesised RNA in the nucleus. *Cell.* **58**: 269-279

Braddock, M., Thorburn, A.M., Chambers, A., Elliott, G.D., Anderson, G.J., Kingsman, A.J., and Kingsman, S.M. (1990) A nuclear translation block imposed by the HIV-1 U3 region is relieved by the TAT-TAR interaction. *Cell.* **62**: 1123-1133

Braddock, M., Thorburn, A.M., Kingsman, A.J., and Kingsman, S.M. (1991) Blocking of TAT dependent HIV-1 RNA modification by an inhibitor of RNA polymerase II processivity. *Nature* **350**: 439-441

Dayton, A.I., Sodroski, J.G., Rosen, C.A., Goh, C.A., and Haseltine, W.A. (1986) The transactivator gene of the human T-cell lymphotropic virus III is required for replication. *Cell* **44**: 941-947

De Wet, J.R., Wood, K.V., DeLuca, M., Helinski, D.R., and Subramani, S. (1987) Firefly luciferase gene: Structure and expression in mammalian cells. *Mol. Cell. Biol.* **7**: 725-737

Gallie, D.R. (1991) The CAP and poly(A) tail function synergistically to regulate mRNA translational efficiency. *Genes Dev.* **5**: 2108-2116

Gallo, R.C., Salahuddin, S.Z., Popovic, M., Shearer, G.M., Kaplan, M., Haynes, B.F., Palker, T.J., Redfield, R., Oleske, J., Safai, B., White, G., Foster, P., and Markham, P.D. (1984) Frequent detection and isolation of cytopathic retroviruses (HTLV-III) from patients with AIDS and at risk for AIDS. *Science* **224**: 500-503

Gorman, C.M., Moffat, L.F., and Howard, B.H. (1982) Recombinant genomes which express chloramphenicol acetyltranferase in mammalian cells. *Mol. Cell. Biol.* **2**: 1044-1051

Gould, S.J., and Subramani, S. (1988) Firefly luciferase as a tool in molecular and cell biology. *Anal. Biochem.* **175**: 5-13

Jakobovits, A., Smith, D.H., Jakobovits, E.B., and Capon, D.J. (1988) A discrete element 3' of the human immunodeficiency virus-1 (HIV-1) and HIV-2 mRNA initiation sites mediates transcriptional activation by an HIV transactivator. *Mol. Cell. Biol.* **8**: 2555-2561

Malone, R.W., Felgner, P.L., and Verma, I.M. (1989) Cationic liposome-mediated RNA transfection. *Proc. Natl. Acad. Sci. USA* **86**: 6077-6081

Maxwell, I.H., and Maxwell, F. (1988) Electroporation of mammalian cells with firefly luciferase expression plasmid: kinetics of transient expression differ markedly among cell types. *DNA* **7**: 557-562

Muesing, M.A., Smith, D.H., and Capon, D.J. (1987) Regulation of mRNA accumulation by a human immunodeficiency virus transactivator protein. *Cell.* **48**: 691-701

Ow, D.W., Wood, K.V., DeLuca, M., De Wet, J.R., Helinski, D.R., and Howell, S.H. (1986) Transient and stable expression of the firefly luciferase gene in plant cells and transgenic plants. *Science* **234**: 856-859

Palomares, A.J., DeLuca, M.A., and Helinski, D.R. (1989) Firefly luciferase as a reporter enzyme for measuring gene expression in vegetative and symbiotic *Rhizobium meliloti* and other Gram negative bacteria. *Gene* **81**: 55-64

Rice, A.P., and Mathews, M.B. (1988) Transcriptional but not translational regulation of HIV-1 by the tat gene product. *Nature* **322**: 551-555

Rosen, C.A., Sodroski, J.G., Goh, W.C., Dayton, A.I., Loppke, J., and Haseltine, W.A. (1986) Post-transcriptional regulation accounts for the trans-activation of the human T-lymphotropic virus type III long terminal repeat. *Nature* **319**: 555-559

Rosen, C.A., Terwilliger, E.L., Dayton, A., Sodroski, J.G., and Haseltine, W.A. (1985) Intragenic cis-acting art gene-responsive sequences of the human immunodeficiency virus. *Proc. Natl. Acad. Sci. USA.* **85**: 2071-2075

Roy, S., Delling, V., Chen, C.H., Rosen, C.A., and Sonnenberg, N. (1990) A bulge structure in HIV-1 TAR RNA is required for TAT binding and TAT-mediated transactivation. *Genes Dev.* **4**: 1365-1373

Thompson, J.F., Hayes, L.S., and Lloyd, D.B. (1991) Modulation of firefly luciferase stability and impact on studies of gene regulation. *Gene* **103**: 171-177

Varmus, H. (1988) Regulation of HIV and HTLV gene expression. *Genes Dev* **2**: 1055-1062

White, M.R.H., Craig, F.F., Watmore, D., McCapra, F., and Simmonds, A.C. (1991) Applications of the direct imaging of firefly luciferase gene expression in mammalian cells, *in:* "Bioluminescence and Chemiluminescence: Current Status", P.E. Stanley and L.J. Kricka, eds., Wiley, Chichester, 357-360

White, M.R.H., Morse, J., Boniszewski, Z.A.M., Mundy, C.R., Brady, M.A.W., and Chiswell, D.J. (1990) Imaging of firefly luciferase expression in single mammalian cells using high sensitivity charge-coupled device cameras. *Technique* **2**: 194-201

APPLICATION OF CONFOCAL MICROSCOPY FOR THE STUDY OF NEURONAL ORGANIZATION IN HUMAN CORTICAL AREAS AFTER MICROINJECTION OF LUCIFER YELLOW

Pavel V. Belichenko[1], Annica Dahlstrom[2], and Patrick Sourander[3]

[1]Brain Research Institute, Russian Academy of Medical Sciences, per. Obukha 5, 103064, Moscow, Russia
[2]Institute of Neurobiology, Department of Histology, and
[3]Department of Pathology, Division of Neuropathology, University of Göteborg, Medicinaregatan 5, S-413 90 Göteborg, Sweden

SUMMARY

The relationship between 2-D or 3-D neuronal morphology and the intrinsic organization of different cortical areas in the human brain was investigated after intracellular microelectrode injection of Lucifer Yellow (the Lucifer Yellow-method). In some areas pyramid and nonpyramidal neurons were differentiated by the presence of autofluorescent lipofuscin granules. The Lucifer Yellow-method allows the visualization in a brain specimen obtained from the neurosurgery clinic of a large number (200-300) of single neurons including the dendrites and axons with collaterals in a small area of tissue. The method has been tested in the study of frontal, temporal, parietal, and occipital cortices, and in the hippocampus in normal and pathological human brain areas. The number of neurons is sufficient to allow all necessary morphometrical investigations needed for each patient. Since neurons are inherently 3-D in their organization, 3-D reconstructions and analysis of neurons using confocal laser scanning microscopy (CLSM) in combination with VoxelView™ software from Vital Images Inc. gives additional information to the classical 2-D representation on neuronal network organization. Data can be demonstrated using 3-D

Biotechnology Applications of Microinjection, Microscopic Imaging, and Fluorescence, Edited by P.H. Bach *et al.*, Plenum Press, New York, 1993

29

pseudocoloured moving reconstructions on video tapes. These new data represent a starting point for contributions from our laboratories to the Human Brain Mapping Project.

INTRODUCTION

The morphological features of individual neurons is a basis of our understanding of normal, as well as of abnormal, functions of the nervous system. Important information on the internal organization of the human cortex was obtained from cytoarchitectural studies that revealed cortical layers and area boundaries (Brodmann, 1909; Braak, 1976; Haug, 1984, and others). In many cytoarchitectural studies the investigation of the distribution of neurons in the cortex was based on the cell size and density (Haug, 1984). Previous studies on the neuronal structure of the human cerebral cortex were focussed on descriptions of dendritic morphology as revealed by the Golgi method (Marin-Padilla, 1969, 1984; Braak, 1976; Poliakov, 1979; Mrzljak *et al.*, 1988). However, camera lucida drawings of Golgi pictures only gives 2-D information about neuronal geometry. Little is known about the principles of cortical network organization within one cortical layer, due to difficulties inherent in the morphological methods used, as well as in the quantitative analysis of neurons in a small area of cortex. However, along with common properties, neurons within a circumscribed area in the same cortical layer also possess morphological and physiological individual differences (Katz, 1987).

The Lucifer Yellow-method clearly has advantages over earlier methods to study individual neurons in the brain of higher mammals:-

Firstly, single neurons can be excellently labelled in all areas and processes by fluorescent contrasting molecules, e.g. Lucifer Yellow, in a sufficiently controlled mode for the staining of human nervous tissue (Buhl and Schlote, 1987; Einstein, 1988; Belichenko and Schlote, 1989; Belichenko, 1991, 1992; Belichenko *et al.* 1992).

Secondly, confocal laser scanning microscopy can form high signal-to-noise ratio optical sections through thick slices and collect the images in digital format (Deitch at al., 1990a,b; Turner at al., 1991).

Thus, Lucifer Yellow-labelling of neurons is ideal for 3-D confocal imaging and constitutes an important new tool for high-resolution 3-D imaging of individual neurons that can be of important assistance in the creation of the Human Brain Mapping project.

The aim of our work was to study the 3-D architecture of individual human cortical neurons in normal and diseased brain tissue by using the Lucifer Yellow-method in combination with CLSM.

MATERIAL AND METHODS

The material used was removed during neurosurgery from 3 patients with therapy resistant epilepsy. Specimens from two temporal cortices (from one 18-year-old male, area 21 in the left posteriotemporal cortex, and one 28-year-old female, area 21 in the right anteriotemporal cortex) and from one occipital cortex (from a 25-year-old female, area 18 in the left parieto-occipital cortex) were removed during operation. The specimens were immediately immersed in cold 4% paraformaldehyde in 0.1 M phosphate buffer (PB) at pH 7.4. Subsequently a small block of the cortex (2 x 1 x 1 mm) was excised, further immersed in the same cold fixative for 4 hour, and then rinsed in PB. The brain tissue was serially sectioned perpendicular to the pial surface with an Oxford vibratome at 150 μm thickness, and stored in PB at 4°C until Lucifer Yellow injection was performed.

For Lucifer Yellow injections, a slice was floated onto a glass plate, and held down by a Millipore™ filter with a slightly smaller window than the underlying tissue. The preparation was immersed in a Petri dish with PB and the injection chamber was then transferred to an Axiomat fluorescence microscope (Zeiss, Germany), placed on a vibration free table. The glass microelectrodes were filled with a 6% aqueous solution of Lucifer Yellow. Electrode resistance varied between 100-150 Mo. Cells could easily be selected for injection due to the presence of autofluorescent lipofuscin particles. Sometimes pyramid neurons and nonpyramidal neurons could easily be distinguished due to the presence of differently distributed lipofuscin granules. Normally, lipofuscin granules in pyramid cells were diffusely localized, while in nonpyramidal neurons the lipofuscin granules formed a more compact mass. After penetration of the microelectrode into the cell body, Lucifer Yellow was iontophoretically injected with a negative constant current of 2-5 nA for 5-10 minutes, until all fine dendritic branches and collaterals of the axon appeared brightly fluorescent. A number of pyramidal and nonpyramidal neurons of layers II-III, V, and the white matter in a number of consecutive slices were iontophoretically injected with the Lucifer Yellow solution. Usually, in each slide it was possible to inject and fill up to 50 neurons. The slices were then mounted on a microscope glass slide and coverslipped with glycerol. Some of the vibratome slides were later stained with the Nissl stain in order to distinguish laminar and area borders, or with immunohistochemical methods, to give further information on e.g. glial elements or transmitter content.

The sections were studied in a BioRad MRC-600™ CLSM equipment, attached to a H Nikon Optiphot FX microscope, employing an argon or an argon/krypton ion laser. Single neurons were imaged using CLSM-generated serial optical sections in adequate incremental intervals, usually 0.5-2.0 μm. The 2-D reconstruction of each neuron was made by using the linear maximal Z-projection program for serial optical images. The 3-D reconstructions and rotations were computed with a Silicon Graphics IRIS work station using a commercially available α-blending programme (Dreben, 1988) as applied on Voxel View - GT™ programme from Vital Images, Fairfield, Inc.

RESULTS

In the present study, we have obtained data concerning the 3-D morphology of pyramidal neurons in layers III, V, and in the white matter of areas 18 and 21. We were particularly interested in the structure and relationships of neurons from areas with dysgenesis, as diagnosed with neuropathological methods.

Lucifer Yellow injected into the soma filled the whole neuron with all its processes, the dendrites with spines (pyramidal neurons) or varicosities (stellate or granular cells), and axons with collaterals (Figure 1a). The optimal conditions for CLSM imaging of Lucifer Yellow-filled neurons were the following: optical sections were scanned at regular increments of 0.5-2 μm; fast speed scanning was employed; each optical section was the result of 10 scans by Kalman filtering, with a division factor of 3; the lens magnification used was x 10 or x 20; zoom factor was 2 or 3. Figure 1b is a Z-series, maximal linear projection, of 32 CLSM optical sections of one Lucifer Yellow labelled pyramidal neuron.

Belichenko *et al.*

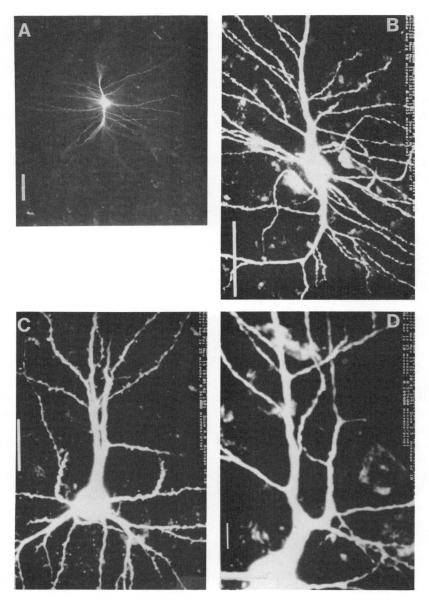

Figure 1. **A)** Lucifer Yellow injected non-pyramidal neuron in human temporal cortex, lamina V, viewed in traditional epifluorescence microscope during the Lucifer Yellow injection procedure. Bar is 50 μm. **B)** Lucifer Yellow injected neuron in human temporal corex, lamina V. The picture is the result of a maximal linear projection of 37 confocal scans. Scale bar is 50 μm. **C)** Lucifer Yellow injected pyramidal neuron in human temporal cortex, lamina III. This neuron has a duplicated apical dendrite, which brances close to its exit from the cell body. The picture is the result of a maximal linear projection of 40 confocal scans. Scale bar is 25 μm. **D)** The apical portion of a pyramidal neuron in human temporal cortex, lamina III. Close to its exit from the cell body the apical dendrite gives off a branch at a right angle from the main direction. The picture is the result of a maximal linear projection of 53 confocal scans. Scale bar is 10 μm. All specimens were from patients with therapy resistant epilepsy. Brain surface at top.

Non-pyramidal neurons and glial cells (not shown) were also filled and imaged using these methods.

We have shown that the Lucifer Yellow method combined with CLSM can contribute significantly to our understanding of the neuronal organization in the cortex, and thereby we may also in the long perspective obtain information about the relationship between structure and function. The preliminary results demonstrate different types of 3-D dendritic abnormalities of single pyramidal neurons in temporal and occipital cortices in laminae III and V. In addition, grossly abnormal pyramidal neurons (with deviating and malformed apical dendrites) were observed in the subcortical layer and in the white matter in some patients with therapy resistant epilepsy (Belichenko et al. 1992). Many cells were found to have two or three dendrites originating from the apical portion, rather than one single apical dendrite (Figure 1a-d).

The advantage of 3-D CLSM imaging of neurons is that the technique enables the reconstruction of the entire neuron with its dendritic and axonal network. Sometimes, only by rotation of 3-D CLSM images, it was possible to distinguish between inverted pyramidal neurons and neurons with two apical dendrites, with one cutting dendrite near the cell body. Due to its sensitivity and the possibility of computer enhancing the signal-to-noise ratio of e.g. Lucifer Yellow fluorescence, CLSM is very suitable for quantitative analysis of the geometry of neurons; a software programme is being developed to be used for an ongoing study utilizing Z-series of CLSM neuronal images.

It should be mentioned that the Lucifer Yellow method in combination with CSLM optical sections, which are automatically registered, and with computer based morphometrical programmes, is likely to open new avenues for research included in the internationally supported 3-D Human Brain Mapping Project.

DISCUSSION

In this study we propose an important and easy strategy for investigating the organization of neuronal networks, a strategy which, hopefully, will become of value for the creation of a data base for the Human Brain Mapping Project.

Firstly, iontophoretic injection of the fluorescent dye Lucifer Yellow into lightly fixed material was successful when used in human biopsy material or tissues removed at autopsy soon after death (Buhl and Schlote, 1987; Einstein, 1988; Belichenko and Schlote, 1989; Belichenko, 1991, 1992; Belichenko et al. 1992). The present paper demonstrates that this method using immersion-fixed human brain material under ordinary biopsy conditions is feasible for normal as well as pathological tissues. Over a period of more than 15 days it was possible to stain cortical neurons in a satisfactory way (Belichenko and Schlote, 1989; Belichenko, 1991,1992). Although these cells were labelled after fixation they closely resemble those cells which are visualized with in vivo labelling techniques, as shown recently (Felthauser and Claiborne, 1990).

Secondly, using 3-D morphological analysis of Lucifer Yellow-filled neurons we found a number of different dendritic abnormalities in pyramidal neurons in layers III and V and in the white matter. This type of work is hardly possible using the Golgi method, because this technique is very unpredictable in its staining, and since it is very difficult to find a large enough number of neurons in a small piece of a Golgi-impregnated slice. Golgi-labelled neurons can also be imaged by CLSM (Tredici et al., 1991), but the Lucifer Yellow-method gives a clearer picture, because we can inject chosen neurons located separately. 3-D CLSM reconstruction of Lucifer Yellow-labelled neurons using the Voxel View-GT™ imaging programme is of great assistance in several cases: when studying parts of basal dendrites and axon close to the cell body; when dendrites are crossing, when it is necessary to calculate real numbers of dendritic spines, or when a glial cell was also labelled together with a neuron. A possible point of contact of crossing dendrites was easily tested using 3-D Voxel View™ rotations.

The distribution of autofluorescent lipofuscin particles within neurons, glial cells, and cells lining the blood vessels in the human brain may, in 2-D pictures, sometimes be the cause of complicated, difficult to interpret, patterns. However, using 3-D CSLM background fluorescence can be reduced or eliminated by computer assisted software. On the other hand, the autofluorescent background structures may be useful in rendering the optical sections more informative about the relation between the Lucifer Yellow injected neuron and surrounding structures.

Recent semiautomatic drawing techniques can produce more accurate 3-D representations of neuronal arbors (Capowsky, 1989) than CSLM. However, the main advantages of CSLM optical sections are:-

a) the automatical registering in computer memory,

b) the high speed which implies that less time is needed to create the picture for each neuron, and

c) the result is a picture which is closer to reality.

These advantages are quite important when many neurons need to be processed in a limited time.

ACKNOWLEDGEMENTS

The authors are grateful to Associate Prof. C. von Essen for providing the human biopsy material, to Associate Prof. C. Nordborg for histopatological diagnosis of the human biopsy material, and to Assoc. Prof. S. Lindström for performing the intracellular injections. The study was supported by research grants from the Swedish MRC (2207), by The Royal Society of Arts and Sciences of Göteborg, and by IngaBritt och Arne Lundbergs Research Foundation.

REFERENCES

Belichenko, P. (1991) Using photooxidation procedure of fluorescent slices of autopsy human brain tissue for light and electron microscopy. *Bulletin Experimental Biology and Medicine* 9: 323-325

Belichenko, P. (1992) Development of an intracellular lucifer yellow method for studying neuronal structures in autopsies of the human brain. *Archiv Anatomy, Histology and Embryology* (in press)

Belichenko, P., Dahlström, A., von Essen, C., Lindström, S., Nordborg, C., and Sourander, P. (1992) Lucifer yellow injected abnormal pyramidal neurons in epileptic human cortex. Confocal laser scanning and 3-D reconstructions. *NeuroReport* (in press)

Belichenko, P., and Schlote, W. (1989) Morphometric analysis of pyramidal neurons in the human occipital and parietal cortex. *Clin. Neuropathology* 5: 220

Braak, H. (1976) On the striate area of the human isocortex. A Golgi and pigmento-architectonic study. *J. Comp. Neurol.* 166: 341-363

Brodmann, K. (1909) Vergleichende Lokalisationslehre der Grosshirnrinde. Barth-Verlag, Leipzig

Buhl, E.H., and Schlote, W. (1987) Intracellular Lucifer Yellow staining and electron microscopy of neurones in slices of fixed epitumourous human cortical tissue. *Acta Neuropathologica* 75: 140

Capowski, J.J. (1989) Computer Techniques In Neuroanatomy, Plenum, New York

Deitch, J.S., Smith, K.L., Lee, C.L., Swann, J.W., and Turner J.N. (1990) Confocal scanning laser microscope images of hippocampal neurons intracellularly labeled with biocytin. *J. Neurosci. Meth.* **33:** 61-67

Deitch, J.S., Smith, K.L., Swann, J.W., and Turner, J.N. (1990) Ultrastructural investigation of neurons identified and localized using the confocal scanning laser microscope *J. Electron Microsc. Tech.* **18:** 82-90

Drebin, A. (1988) 3D Volume rendering, Computer graphics, *Proc. SIGGRAPH '88* **22:** 65-74

Einstein, G. (1988) Intracellular injection of Lucifer Yellow into cortical neurons in lightly fixed sections and its application to human autopsy material. *J. Neurosci. Meth.* **26:** 95-179

Felthauser, A.M., and Claiborne, B.J. (1990) Intracellular labeling of dentate granule cells in fixed tissue permits quantitative analysis of dendritic morphology. *Neurosci. Lett.* **118:** 249-251

Haug, H. (1984) Macroscopic and microscopic morphometry of the human brain and cortex. A survey in the light of new results. *Brain Pathol* **1:** 123-149

Marin-Padilla, M. (1969) Origin of the pericellular baskets of the pyramidal cells of the human motor cortex: A Golgi study. *Brain Res.* **14:** 633-646

Marin-Padilla, M. (1984) Neurons of layer I. A developmental analysis, in: "Cerebral Cortex, Vol. 1: Cellular Components of The Cerebral Cortex", A. Peters and E.G. Jones, eds., Plenum Press, New York, 447-478

Mrzljak, L., Uylings, H.B.M., Kostovic, I., and Van Eden, C.G. (1988) Prenatal development of neurons in the human prefrontal cortex: A Qualitative Golgi study. *J. Comp. Neurol.* **271:** 355-386

Poliakov, G.I. (1979) Entwicklung Der Neuronen Der Menschlichen Gross hirnrinde, VEB Georg Thieme, Leipzig

Tredici, G., Di Francesco, A., Cavaletti, G., Pizzini, G., and Miami, A. (1991) 3-Dimentional images of Golgi impregnated and HRP labeled neurons by confocal scanning microscope, *in:* "Abstract at the Third IBRO Congress of Neurosciences," 55

Turner, J.N., Szarowski, D.H., Smith, K.L., Marko, M., Leith, A., and Swann, J.W. (1991) Confocal Microscopy and three-dimentional reconstruction of electrophysiologically identified neurons in thick brain slices. *J. Electron Microsc. Tech.* **18:** 11-23

THE DYE-TRANSFER ASSAY PERMITS THE EVALUATION OF THE MODULATION OF JUNCTIONAL COMMUNICATION BY CHEMICALS: RESULTS OBTAINED EMPLOYING A HIGHLY SENSITIVE VIDEO-RECORDING SYSTEM CONNECTED WITH THE MICROINJECTOR

Giovanna Mazzoleni[1], Anna Camplani[1], Paola Telò[1], Silvia Tanganelli[1], and Giovanni Ragnotti[2]

[1]Unit of General Pathology and Immunology, Department of Biomedical Science and Biotechnology, School of Medicine, University of Brescia, Via Valsabbina 19, 25123 Brescia, Italy

[2]Chair of General Pathology, School of Medicine, University of Milan, Via Festa del Perdono 7, 20122 Milano, Italy

INTRODUCTION

Gap-junctional intercellular communication (GJIC) plays a crucial role in the regulation of cell proliferation and differentiation as well as in the maintenance of tissue homeostasis (Lowenstein, 1979; Gilula, 1977). Since the control of these functions is disrupted when the cells become neoplastic, it has been proposed that GJIC modulation is involved in multistage carcinogenesis (Gilula, 1977; Weinstein *et al.*, 1976). In particular it has been postulated that the inhibition of GJIC is a pivotal event in the promotion phase of carcinogenesis, in that it enables the initiated cell to isolate itself from the suppressive influence of normal neighboring cells thus allowing the clonal expansion to form a critical pre-neoplastic mass (Trosko *et al.*, 1982; Yamasaki *et al.*, 1984).

On this basis, the use of GJIC assay has been proposed to detect the tumour-promoting activity of environmental chemicals, since no viable *in vitro* short-term test is currently available for this purpose (Zeilmaker *et al.*, 1986).

Biotechnology Applications of Microinjection, Microscopic Imaging, and Fluorescence, Edited by P.H. Bach *et al.*, Plenum Press, New York, 1993

Several techniques (i.e. metabolic cooperation, electrical coupling, dye-injection, scrape loading/dye-transfer and fluorescence recovery after photobleaching) are available to evaluate the GJIC in cultured cells and its modulation by chemicals. Unlike other methods, the dye-transfer assay (dye-injection and scrape loading technique) facilitates the study of the junctional communication in cells cultured under any experimental condition and it better reflects their physiological status.

To determine gap-junctional permeability by the microinjection technique, individual cells in monolayer cultures are microinjected with a junctional permeant fluorescent tracer (Lucifer Yellow CH) and the extent of GJIC is estimated by the number of fluorescent neighboring cells scored within 10 minutes from the injection.

In this study the modulation of GJIC by several chemicals known to be both tumour promoting agents (i.e. 12-O-tetradecanoylphorbol-13-acetate, TPA; phenobarbital, PB; phorbol-12,13-diacetate, PDA and 1,1,1-trichloro-2,2-bis(p-chlorophenyl)ethane, DDT) (Boutwell, 1978; Peraino *et al.*, 1975; Trosko *et al.*, 1982) and non-tumour promoting agents (i.e. 4α-phorbol-12,13-didecanoate, 4α-PDD and phorbol, PHR) (Trosko *et al.*, 1982) is evaluated by the microinjection/dye-transfer technique in the endothelial cell line F-BAE GM7373 (Grinspan *et al.*, 1983).

A direct microscopic determination of the number of communicating cells presents some difficulties because of:

a) the need to measure the dye-spreading area rapidly, due to the decay of the fluorescence;

b) the impossibility of marking the loaded cells, excluding those already checked, and

c) the subjectivity of the quantification of the results, thus distorting the evaluation of the data.

To overcome these restrictions we propose the use of a highly sensitive video recording system. The system is composed of a charge coupler device (CCD) videocamera connected to the microinjector; the camera leads the images to a monitor and to a videorecorder, allowing the recording of the microscopic images and permitting a delayed evaluation of the results. The microscopic images can be printed with a video printer connected to the microscope and to the recorder; the recorded experiment can eventually be filed for further analysis.

MATERIALS AND METHODS

Chemicals and Materials

Lucifer Yellow CH, TPA, 4α-PDD, PDA, PHR, PB, and DDT were purchased from Sigma Chemical Co., St. Louis, MO. All other chemicals were of analytical grade. TPA, PHR, PDA, 4α-PDD, PB and DDT were dissolved in DMSO. Exposure to chemicals was carried out for time periods ranging from 4 to 48 hours. The capillary tubes used for microinjection were from A-M Systems Inc., Everett, WA.

Cell Cultures

F-BAE GM7373 cells were obtained from the National Institute of General Medical Sciences (N.I.G.M.S.) Human Genetic Mutant Cell Repository (Institute for Medical Research, Camden, NJ). They represent an immortalized cell line derived from normal

F-BAE cells by *in vitro* transformation with benzo(a)pyrene and correspond to the BFA 1c-1BTP multilayered transformed clone described by Grinspan *et al.*, (1983). The endothelial cell line used in this study was maintained in Eagle-Minimal Essential Medium (E-MEM) supplemented with 10% foetal calf serum, essential and non-essential amino acids, antibiotics and vitamins and grown in a 5% CO_2 humidified atmosphere at 37°C. Stock cultures were maintained by plating 1×10^6 cells/100 mm-dish and subculturing them twice a week.

For experimental use, GM7373 in logarithmic growth phase were treated with trypsin-EDTA and plated in the presence of 10% foetal calf serum supplemented medium, at a density of 5×10^5 cells/60 mm-dish, in order to obtain a confluent monolayer 72 hours after the seeding (Figure 1). Twenty-four hours after the plating, the cultures were supplemented every 24 hours for 48 hours with fresh medium containing 10% foetal calf serum and the chemical to be tested or the control solvent (DMSO). The final concentration of the solvent in the culture medium was less than 0.25%(v/v); at this concentration DMSO had no effect either on the growth rate or on the GJIC capacity of the endothelial cells.

Microinjection Technique

Cell-to-cell coupling was evaluated by the microinjection/dye-transfer technique. The intercellular transfer of the gap junction-permeant tracer Lucifer Yellow CH (Stewart, 1981) was estimated after direct microinjection of the dye into the cells. A 10% (w/v)

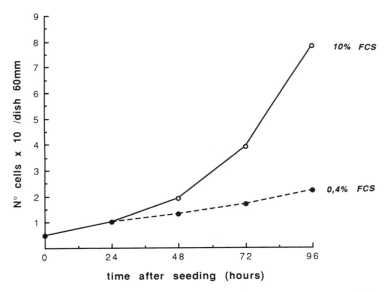

Figure 1. In the presence of 10% foetal calf serum the growth of GM7373 cells shows the exponential pattern typical for a transformed cell line while GM7373 cultures maintained in 0.4% foetal calf serum grow until they reach a confluent monolayer, this being indicative of the strict serum-dependence of the proliferative capacity of these cells.

solution of Lucifer Yellow CH in 0.33 M lithium chloride was transferred to a glass needle prepared from a capillary tube using a dual step puller (Narishige, Japan).

Individual cells in confluent areas were impaled with the needle close to the nucleus and the dye was injected continuously under nitrogen gas pressure, using an Eppendorf 5242 microinjector (Hamburg, Germany) connected to the Olympus microinjectoscope IMT-2 SYF-II equipped with reflected light fluorescence attachment IMT-2 RFL-1 (Mesnil

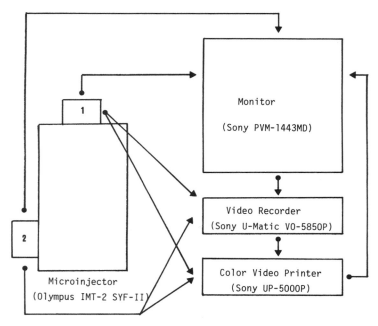

Figure 2. Schematic representation of the connections between the microinjectoscope and the elements of the video-recording system employed for the quantification of the GJIC dye-transfer assay. Using a videocamera (Sony 3 CCD DX-750P) and an image intensifier (Hamamatsu C2400).

et al., 1988). During the manipulation period cell cultures were maintained at room temperature. The number of the dye-coupled cells/injections was used as an index of the extent of GJIC. More than 20 injections were scored per sample and each experimental point represents the mean ± S.E.M. of two independent experiments.

The extent of the dye transfer was estimated by scoring the dye-coupled cells/injections on recorded images obtained with a highly sensitive video-recording system connected with the microinjectoscope (Figure 2). This system is composed of an Olympus IMT-2 microinjector connected with a Sony 3 CCD DX-750 P videocamera that drives the microscopic images to a Sony PVM-1443MD monitor or to a Sony U-MATIC VO-5850P video recorder. The recorded microscopic images can be visualized on the monitor for a delayed quantification of the results and, if necessary, they can be printed with a Sony UP-5000P colour videoprinter.

RESULTS

Characterization of Junctional Communication in F-BAE GM7373 Cells and Inhibition by TPA

Before studying the effect of the various chemicals on intercellular communication capacity in F-BAE GM7373 cells, we characterized the junctional communication in GM7373 cells untreated and treated with TPA as positive control (manuscript in preparation).

Communication competence of GM7373 cells was evaluated by the dye-transfer technique. GM7373 cells showed (Figure 3) a very high communication capacity when maintained in basal conditions (more than 50 dye-coupled cells/injection). The extent of GJIC was related to the cellular density, being maximal in a confluent monolayer. Serum, while affecting the GM7373 growth rate (Figure 1), did not significantly influence the extent of GM7373 dye-coupling.

Before being assayed for its capacity to affect GJIC, TPA was tested for its cytotoxicity, considering the inhibition of endothelial cell growth rate as an index of toxicity. Thus, endothelial cell cultures were treated with increasing concentrations of TPA.

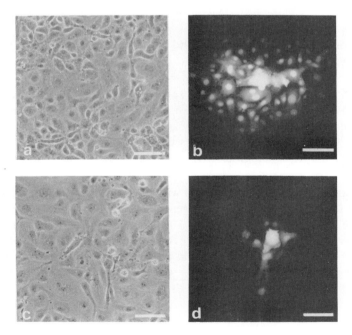

Figure 3. (a-b) Control F-BAE GM7373 cells show a high basal GJIC capacity. **(c-d)** TPA-treated cells. A 24-hour treatment with 100 ng/ml TPA drastically reduces the endothelial GJIC capacity. Bar = 150 μm.

Table 1. Communication capacity of F-BAE GM7373 (% of the control $\pm$ S.E.M.).

Chemical	Concentration	Time of Treatment (hours)		
		4	24	48
	10 ng/ml	56.5 ± 16.4	6.0 ± 1.0	6.5 ± 4.2
TPA	30 ng/ml	24.6 ± 4.9	13.2 ± 1.3	-
	100 ng/ml	31.7 ± 1.9	8.3 ± 1.0	8.4 ± 0.7
PB	10 μM	72.0 ± 2.6	82.3 ± 3.6	-
PDA	100 ng/ml	-	65.6 ± 4.8	-
DDT	5 μM	89.2 ± 4.3	96.1 ± 3.2	-
PHR	100 ng/ml	-	88.9 ± 6.1	-
4-PDD	100 ng/ml	-	105.6 ± 3.3	-

TPA effect on cellular growth rate was evaluated 24 and 48 hours later. In order to avoid non-specific effects on the dye-transfer assay the chemicals were used at non-cytotoxic doses, that is doses that did not negatively affect cell growth rate.

TPA is a potent inhibitor of junctional communication in GM7373 (Table 1). At non-cytotoxic, mitogenic concentrations (100 ng/ml) TPA significantly decreases GM7373 junctional communication in a dose- and time-dependent manner during the 48 hour period of treatment (Figure 3); 10% foetal calf serum did not modify these results.

Effects of Some Tumour-Promoting and Non Tumour-Promoting Agents on Dye-Transfer between GM7373 Cells

GM7373 cells were exposed for 24 hours to the phorbol esters PDA (a weak promoter), PHR or 4α-PDD (both inactive as tumour promoters) at the concentrations shown to be non-cytotoxic with the toxicity assay described above. No significant effect on cell-coupling was observed after 24 hours exposure to 100 ng/ml PHR and 100 ng/ml 4α-PDD, while a 35% inhibition of GJIC was found after treatment with 100 ng/ml PDA (Table 1).

Further, GM7373 cells were treated with the highest non-cytotoxic concentration of the powerful rat liver tumour-promoting agents PB and DDT (Peraino *et al.*, 1975). When the cells were incubated in the presence of 10 μM PB, a significant inhibition of GJIC was observed after 4 hours of treatment (Table 1); the same degree of inhibition (30%) was obtained after 1 hour of exposure to 5 μM DDT (data not shown). In both cases, GJIC capacity was partially (PB) or completely (DDT) recovered within 24 hours from the beginning of the treatment.

DISCUSSION

We have shown by the dye-transfer technique that the tumour promoters TPA, PB, PDA and DDT inhibit intercellular communication between cultured F-BAE GM7373 cells, while no significant effect on GJIC capacity is observed after exposure to the non-tumour promoters PHR and 4α-PDD. After 4 hours treatment with TPA, PB, DDT and 24 hours treatment with PDA, a significant inhibition of GJIC was observed, while cells treated with PHR and 4α-PDD mimicked the controls.

The results of the dye-transfer assay were obtained by scoring the dye-coupled cells/microinjection on recorded microscopic images displayed on the monitor of a high sensitive video-recording system connected to the microinjection apparatus. This device allows us to overcome the difficulties encountered during the direct microscopic quantification of the dye spreading (see Introduction). It also facilitates storage of the data for subsequent analysis and the filing or archiving of the experiments if desired.

Since the inhibition of intercellular communication is considered closely associated with tumour promotion, the results obtained indicate that the endothelial cell line used in this study represents a tumour promoter-sensitive system. Furthermore, the dye-transfer assay allows, in cultured cells, the functional study *in vitro* of GJIC capacity and its modulation by chemicals.

ACKNOWLEDGEMENTS

This work was supported by grants from E.E.C. (BIOT-CT91-0261) and Italcementi S.p.A. to G.R.

REFERENCES

Boutwell, R.K. (1978) Biochemical mechanism of tumor promotion, *in:* "Mechanism of Tumor Promotion and Cocarcinogenesis." T.J. Slaga, A. Sivak and R.K. Boutwell, eds., **2**: 49-58

Gilula, N.B. (1977) Gap junctions and cell communication, *in:* "International Cell Biology." B. Brinkly and K.R. Potter, eds., New York Rockfeller University Press, New York, 61-69

Grinspan, J.B., Mueller, S.N., and Levine, E.M. (1983) Bovine endothelial cells transformed *in vitro* by benzo(a)pyrene. *J. Cell Physiol.* **114**: 328-338.

Loewenstein, W.R. (1979) Junctional intercellular communication and the control of growth. *Biochim. Biophys. Acta.* **560**: 1-65

Mesnil, M., and Yamasaki, H. (1988) Selective gap-junctional communication capacity of transformed and non-transformed rat liver epithelial cell lines. *Carcinogenesis* **9**: 1499-1502

Peraino, C., Michael, F., Staffeldt, E., and Christopher, J.P. (1975) Comparative enhancing effects of phenobarbital, amobarbital, diphenylhydrantoin and dichlorodiphenyltrichloroetane on 2-acetyl-aminofluorene-induced hepatic tumorigenesis in the rat. *Cancer Res.* **35**: 2884-2890

Stewart, W.W. (1981) Lucifer dyes-highly fluorescent dyes for biological tracing. *Nature* **292**: 17

Trosko, J.E., Yotti, L.P., Warren, S.T., Tsushimoto, G., and Chang, C.C. (1982) Inhibition of cell-cell communication by tumor promoters. *Carcinog. Compr. Surv.* **7**: 565-585

Weinstein, R.S., Merk, F.B., and Alroy, J. (1976) The structure and function of intercellular junctions in cancer. *Adv. Cancer Res.* **23**: 23-89

Yamasaki, H., Enomoto, T., and Martel, N. (1984) Intercellular communication, cell differentiation and tumor promotion, *in:* "Models, Mechanisms and Etiology of Tumor Promotion", IARC Scientific Publications, M. Borzsonyi, K. Lapis, N.E. Day and H. Yamasaki, eds., International Agency for Research on Cancer, Lyon, France, **56**: 217-238

Zeilmaker, M.J., and Yamasaki, H. (1986) Inhibition of junctional intercellular communication as a possible short-term test to detect tumor promoting agents: results with nine chemicals tested by dye-transfer assay in Chinese Hamster V79 cells. *Cancer Res*, **46**: 6180- 6186

COMPARISON OF FLUORESCENT DYE TRANSFER IN INTERCELLULAR COMMUNICATION STUDIES USING SCRAPE-LOADING AND MICROINJECTION TECHNIQUES

E. Honikman-Leban and M.M. Shahin

Department of Chemical Protection and Photobiological Research *In Vitro*, L'Oreal Advanced Research Centre, 1 Ave. E. Schueller, 93600, Aulnay-sous-Bois, France

INTRODUCTION

Intercellular communication through gap junctions is vital for multicellular organisms. It controls and regulates the exchange of information, nutrients and energy between cells and participates in cell growth and differentiation. Gap junctions are well-defined structures (Loewenstein, 1979, 1981) and they allow cell-to-cell transfer of molecules both in mammalian cells *in situ* and in confluent cells in culture. The gap junction is unique among the epithelial junctional complexes in its function as an organelle involved in communication. It is formed by a collection of membrane proteins (connexons) which interact across the space between neighbouring cells to form continuous cell-to-cell pathways for ions and small molecules (Unwin, 1986). The connexon is composed of 6 identical subunits arranged symmetrically in the plane of the membrane and delimiting the channel along their common axis.

Inhibition of intercellular communication is postulated to be a factor in tumour promotion and other malfunctions. In 1979, Trosko and his group found that the potent promoter TPA (12-0-tetradecanoyl phorbol-13-acetate), used in two-step carcinogenesis experiments on mouse skin, inhibits the gap junctional intercellular communication (metabolic cooperation assay) (Yotti *et al*, 1979).

At the cellular level, inhibition of gap-junctional intercellular communication may lead to the metabolic isolation of some cells, allowing their growth as independent clones. The working hypothesis is based on the concept that interruption of intercellular communication promotes the phenotypic expression of some genetic alteration of the cell.

Biotechnology Applications of Microinjection, Microscopic Imaging, and Fluorescence, Edited by P.H. Bach *et al.*, Plenum Press, New York, 1993

Thus, a relationship is drawn between tumour promoters and inhibition of gap-junctional intercellular communication.

MATERIALS AND METHODS

The intercellular communication is evaluated by the transfer of a vital fluorescent non-permeable dye, which is directly introduced into the cell: diffusion of the dye to the surrounding cells is followed through gap junctions. Two methodologies are used in this dye transfer assay: "scrape-loading", described by El-Fouly *et al.* (1987), and "microinjection", described by Yamasaki *et al.* (1985). For both protocols, 500,000 cells in 5 ml medium are plated in Petri dishes and cultivated as a monolayer to a subconfluent state for 48 hours. After the 48 hours, cells are treated with non-cytotoxic doses of the tested compound for 1, 4, 24 and 48 hours.

For scrape-loading, the medium of the treated cells is discarded and the cells are washed with phosphate buffer saline (PBS) solution. The fluorescent dye, Lucifer Yellow CH is spread over the cell culture (2 ml of 0.5% Lucifer Yellow in 0.33M lithium chloride). At this concentration, the dye is not cytotoxic and does not affect the cell membrane. Immediately, the culture is carefully scraped with the edge of a sharp scalpel, so that the border of the cut is clear and the cell culture keeps its integrity as a monolayer in the dish. The scraped culture is incubated for 2 minutes with the dye solution. This solution is then discarded and the cells are gently washed 3-4 times with PBS before fixation of the dye with a formaldehyde solution (4%). There is no significant difference in the diffusion of the dye whether the scraping is performed before or during the dye incubation (McKarns and Doolittle, 1992). The cells are then ready for microscopic observation. The microscope (Leitz) is equipped for epifluorescence with a 100 W mercury lamp and the detection of the Lucifer Yellow spectrum is made through filter selection (E3 Leitz filter-430/490 nm).

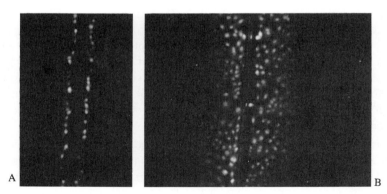

Figure 1. Images of scrape-loading and dye transfer of Lucifer Yellow in a monolayer of untreated (**A**) and TPA (100ng/ml, 1 hour) treated (**B**) IAR20 cells. Intercellular communication was determined by measuring the gradient dye diffusion among the cells, following 2 minutes exposure to 0.5% Lucifer Yellow at room temperature. Visual observation indicates that variation in the dye spreading exists among untreated cells, and that the inhibition observed in the TPA treated cells is significant. No diffusion of the dye is perceptible in the treated cells where the fluorescence is only detectable along the scraped line. See also colour plate 1.

For microinjection, the medium of the treated cells is discarded and replaced with fresh medium. The Petri dish is placed under the fluorescent microscope, and the microinjection is monitored into a single cell with the help of an automatic micromanipulator (Eppendorf) which conducts a microcapillary needle. This needle is prepared from a glass capillary tube (Kwik-Fill, Clarks, with filament) by an automatic puller and filled with a drop of Lucifer Yellow (5% in LiCl solution). The tip of the needle is about 1 μm. The dye is automatically injected into the cell by the microinjector (Eppendorf) which delivers a standardized pulse under nitrogen. Fifteen to twenty microinjections can be performed in a Petri dish during a period of less than 20 minutes. The spreading of the dye is maximum after 5 minutes, and can be evaluated immediately thereafter under microscopic observation or after fixation of the culture in formaldehyde solution; it can also be scored photographically.

All manipulations with the fluorescent dye are conducted under inactinic light in order to avoid photobleaching of the dye and/or fading of the fluorescence intensity.

In our studies we used a rat liver epithelial cell line (IAR20) which was developed by Mesnil *et al.* (1987). The cells were grown in William's E medium supplemented with L-glutamine (1 mM), 100 μg/ml streptomycin and 100 IU/ml penicillin (Eurobio, France) and 10% of heat-decomplemented foetal calf serum (Gibco). Plates were then incubated at 37°C in a humidified atmosphere containing about 5% CO_2 and 95% air. The test compounds were purchased from Sigma Chemical Co, U.S.A. They were dissolved in dimethyl sulfoxide (DMSO, Merck) to give a final concentration of 0.2%.

RESULTS

Results for investigated compounds are generally compared with negative controls (untreated cells, solvent control), diffusion of the dye Lucifer Yellow, and positive control (TPA), inhibition of the dye transfer as shown in Figure 1 (scrape-loading) and Figure 2 (microinjection).

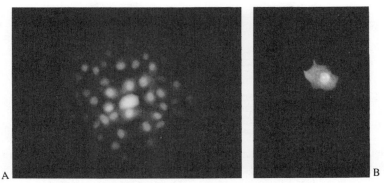

Figure 2. Images of microinjection and dye transfer of Lucifer Yellow in a monolayer of untreated (**A**), and TPA (100 ng/ml, 1 hour) treated (**B**) IAR20 cells. Intercellular communication was determined by counting the cells that became fluorescent after the dye diffused from the microinjected cell, following 5 minutes exposure to 5% Lucifer Yellow at room temperature. See also colour plate 1.

Figure 1a shows the effect of TPA treatment on dye transfer using scrape-loading assay. As a result of TPA cell treatment (100 ng/ml, 1 hour treatment), LY was not transferred into surrounding cells. TPA inhibited dye transfer by more than 90%. In contrast, in control cultures (solvent control) no significant change in the extent of LY transfer was observed (Figure 1b) as shown in plate 1.

Figure 2 shows results after fluorescent Lucifer Yellow was microinjected into a cell of an untreated culture. Immediately, the dye was transferred into the adjacent cells through the gap junctions (Figure 2a). This was not the case when the cultures were treated with TPA. TPA strongly inhibited the transfer of Lucifer Yellow into the surrounding cells, indicating interruption of cell-to-cell communication (Figure 2b). Our present study showed that TPA inhibits the transfer of Lucifer Yellow independent of the methodologies used (scrape-loading and microinjection assay systems).

Quantification is possible with this technique. In the TPA treated cells, only the microinjected cell is fluorescent; the intensity of the fluorescence remains very high in the case of inhibition of gap-junctional intercellular communication, or at least higher than in the untreated cells where the diffusion lowered the fluorescence level and gives an acceleration to the process of fading.

Figure 3 shows the capacity of TPA to inhibit gap-junctional intercellular communication. At 1 ng/ml and 1 hour treatment time, the inhibitory effect on dye transfer was found to be slightly more than 50%. An almost total inhibition was obtained at 10

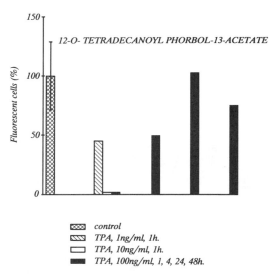

Figure 3. Gap junctional intercellular communication in IAR20 cells as determined after microinjection. The dye transfer was quantified by counting the number of fluorescent cells and the results of the TPA treatment are expressed in percentage over the diffusion control (untreated cells). Monolayer cultures were exposed to TPA: 1, 10, 100 ng/ml for 1 hour, and to TPA: 100 ng/ml for 4, 24 and 48 hours. Sixteen to twenty areas (2 plates, eight to ten random areas per plate) were analyzed for each untreated solvent control group and for each TPA treated group.

ng/ml and 100 ng/ml and one hour treatment time. At 100 ng/ml reversible inhibition at 4, 24, and 48 hours was observed, indicating that TPA concentrations are associated with the expression of dye transfer. This system is now being employed to study the mechanism(s) by which TPA and related tumour promoters inhibit intercellular communication.

Data for 4-phorbol-12,13-didecanoate (4-PDD) are shown in Figure 4. In spite of the high doses and long treatment periods used, all values are similar to those obtained in the solvent controls. The results indicate that this non-promoter phorbol ester, although it was tested under the same experimental conditions as TPA, did not inhibit dye transfer. All controls functioned as expected, strengthening the reliability of the system.

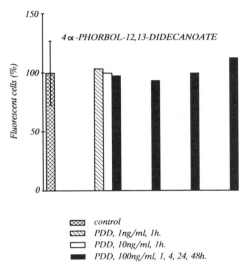

Figure 4. Gap junctional intercellular communication in IAR20 cells as determined after microinjection. The dye transfer was quantified by counting the number of fluorescent cells and the results of the 4-PDD treatment are expressed in percentages over the control (untreated cells). Monolayer cultures were exposed to 4-PDD: 1, 10, 100 ng/ml for 1 hour, and to 4-PDD: 100 ng/ml for 4, 24 and 48 hours. Sixteen to twenty areas (2 plates, eight to ten random areas per plate) were analyzed for each untreated solvent control group and for each 4-PDD treated group.

We also investigated the influence of butylated hydroxyanisole (BHA) on transfer of Lucifer Yellow using the microinjection technique (Figure 5). In contrast to the lack of activity of 4-PDD, after treatment of the cells with 30 µg/ml of BHA for 1, 4, 24 and 48 hours, inhibition of dye transfer was observed. The yields of inhibition were in the range of 75-84%, with very slight variation (Figure 5). On the other hand, 3 and 10 µg/ml after treatment of 1 hour had no effect on cell to cell dye transfer.

In general, no discrepancies have been observed with the chemicals tested (Table 1). The data obtained with the metabolic cooperation as well as with the microinjection techniques are comparable and the systems employed seem to be appropriate for the screening of molecules with a potential promoting activity.

Table 1. Effects on intercellular communication of selected chemicals tested with the dye transfer methodologies (scrape-loading and microinjection) and compared with the metabolic cooperation protocol (Yotti *et al*, 1979).

	Metabolic Cooperation V79 cells	Fluorescent Dye Transfer IAR20 cells	
		Scrape loading	Microinjection
Compounds			
TPA	+++	+++	+++
4α-PDD	-	-	-
BHA	++	+	++
Retinoic acid	++	++	+
Ascorbic acid	-	-	-
α-Tocopherol	-	-	-
DMSO	-	-	-

For non-cytotoxic concentration range: Inhibition of GJIC with dose response effect (+++ to +) and no effect (-)

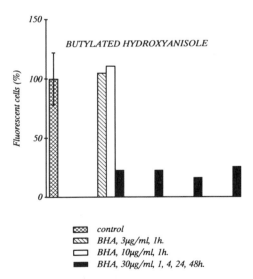

control
BHA, 3µg/ml, 1h.
BHA, 10µg/ml, 1h.
BHA, 30µg/ml, 1, 4, 24, 48h.

Figure 5. Gap junctional intercellular communication in IAR20 cells as determined after microinjection. The dye transfer was quantified by counting the number of fluorescent cells and the results of the BHA treatment are expressed in percentages over the control (untreated cells). Monolayer cultures were exposed to BHA at the concentration of 3, 10, 30 µg/ml for 1 hour, and to BHA 30 µg/ml for 4, 24 and 48 hours. Sixteen to twenty areas (2 plates, eight to ten random areas per plate) were analyzed for each untreated solvent control group and for each BHA treated group.

DISCUSSION

The scrape-loading and microinjection techniques are rapid and easy to handle. Both techniques are being used to test for tumour promoters. The microinjection technique is not only able to identify the compounds as active or inactive but also to indicate the degree of their activity (slight, moderate, strong) and to illustrate certain actions (reversibility) or characteristics (stability) during the contact period. It is also suitable in revealing an enhancement in dye transfer, indicative of an antipromoting effect.

Our findings are in agreement with previous observations (Trosko *et al.*, 1982; Telang *et al.*, 1982) that numerous tumour promoters possess the property of inhibiting cell-to-cell communication, although some exceptions have been noted (Kinsella, 1982). Exceptions are not unusual in the systematic screening of drugs. Short-term tests differ in their pitfalls and none of them is perfect.

However, within the BRIDGE programme (BIOT-91-0261) of the European Community, seven laboratories are collaborating on the development of optimal conditions for the detection of chemicals interfering with gap-junctional intercellular communication (inhibitors and enhancers), using human and animal cell cultures. The mechanisms of gap-junctional intercellular communication and its role during the promotion phase of the carcinogenesis process is also being studied in this European project.

ACKNOWLEDGMENTS

This research has been supported by the BRIDGE programme (BIOT-91-0261) of the European Community.

REFERENCES

El-Fouly, M., Trosko, J.E., and Chang, C.C. (1987) Scrape-loading and dye transfer - a rapid and simple technique to study gap junctional intercellular communication. *Exp. Cell Res.* **168**: 422-430

Kinsella, A.R. (1982) Elimination of metabolic cooperation and induction of sister-chromatid exchanges are not properties common to all promoting or co-carcinogenic agents. *Carcinogenesis* **3**: 499-503

Loewenstein, W.R. (1979) Junctional intercellular communication and the control of growth. *Biochim. Biophys. Acta* **560**: 1-65

Loewenstein, W.R. (1981) Junctional intercellular communication. The cell-to-cell membrane channel. *Physiol. Rev.* **61**: 829-913

McKarns, S.C., and Doolittle, D.J. (1992) Limitations of the scrape-loading/dye transfer technique to quantify inhibition of gap junctional intercellular communications. *Cell Biol. and Toxicol.* **8**: 89-103

Mesnil, M., Fraslin, J.M., Piccoli, C., Yamasaki, H., and Guguen-Guillouzo, C. (1987) Cell contact but not junctional communication (dye coupling) with biliary epithelial cells is required for hepatocytes to maintain differentiated functions. *Exp. Cell Res.* **173**: 524-533

Telang, S., Tong, G., and Williams, G.M. (1982) Epigenic membrane effects of a possible tumour promoting type of cultured liver cells by the non-genotoxic organochloride pesticides, chlordane and heptachlore. *Carcinogenesis* **3**: 1175-1178

Trosko, J.E., Yotti, L., Warren, S.T., Tsushimoto, G., and Chang, C.C. (1982) Inhibition of cell-cell communication by tumour promoters, *in:* "Cocarcinogenesis and Biological Effects of Tumour Promoters," E. Hecker, N.E. Fusening, W. Kuntz, F. Marks and H.W. Thielman, eds., Raven Press, New York, 565-585

Unwin, P.N.T. (1986) Gap junctional structure and control of cell-to-cell communication, *in:* "Junctional Complexes of Epithelial Cells," CIBA Foundation Symposium, G. Bock and S. Clark, eds., Wiley, U.K., **125**: 78-87

Yamasaki, H., Aguelon-Pegouries, A.M., Enomoto, T., Martel, N., Furstenberger, G., and Marks, F. (1985) Comparative effects of a complete tumour promoter, TPA, and a second-stage tumour promoter, RPA, on intercellular communication, cell differentiation and cell transformation. *Carcinogenesis* **6:** 1173-1179

Yotti, L., Chang, C.C., and Trosko, J.E. (1979) Elimination of metabolic cooperation in Chinese hamster cells by a tumour promoter. *Science* **206:** 1089-1091

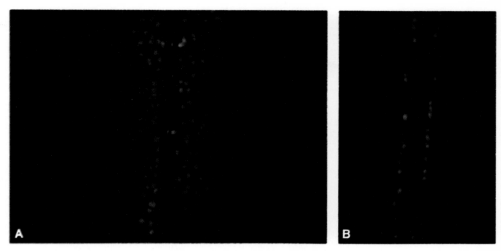

Honikman-Leban, Figure 1: Images of scrape-loading and dye transfer of Lucifer Yellow in a monolayer of untreated (**A**) and TPA (100 ng/ml, 1 hour) treated (**B**) IAR20 cells. Intercellular communication was determined by measuring the gradient dye diffusion among the cells, following 2 minutes exposure to 0.5% Lucifer Yellow at room temperature. Visual observation indicates that variation in the dye spreading exists among the untreated cells, and that the inhibition observed in the TPA treated cells is significant. No diffusion of the dye is perceptible in the treated cells where the fluorescent is only detectable along the scraped line.

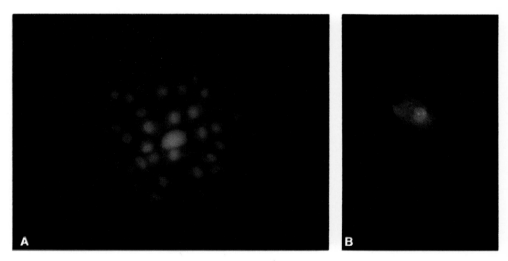

Honikman-Leban, Figure 2: Images of microinjection and dye of Lucifer Yellow in a monolayer of untreated (**A**) and TPA (100 ng/ml, 1 hour) treated (**B**) IAR20 cells. Intercellular communication was determined by counting the cells that became fluorescent after the dye diffused from the microinjected cell, following 5 minutes exposure to 5% Lucifer Yellow at room temperature.

Plate 1

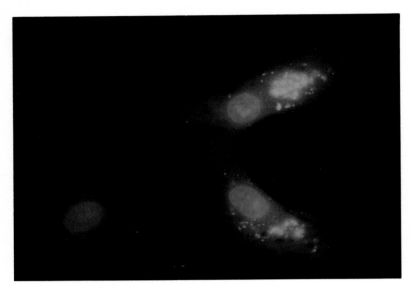

Rashid-Doubell: The fluorescence micrograph shows two cultured fibroblasts obtained from eighteen day old Wistar rat embryos which have had the fluorescent probe Acridine Orange (probe concentration = 0.1% w/v) administered for three minutes and observed using the fluorescein filter. The Acridine Orange probe displays a classic metachromatic pattern of a green nucleus and red lysosomal staining.

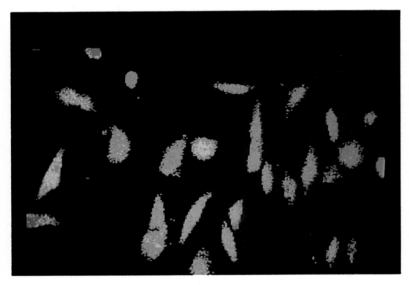

Wiltink et al.: Ratio image recorded with the use of microfluorescence techniques using Fura-2, on foetal rat osteoblasts, illustrating the effect of bleaching on R_{max} during *in vivo* calibration at the end of the experiment. After bleaching the dye, the culture dish containing the osteoblasts was deliberately displaced over part of the field of observation directly before the addition of 2.5 μM ionomycin. The cells at the top and left hand of the image are unexposed cells and reach much higher R_{max} values (mean, R = 2) than the cells in the bleached region (right-hand side; mean, R = 0.8). Pseudocolours are such that red corresponds to high ratio values (2.6); blue corresponds to low ratio values (0.1).

Plate 2

CHARACTERIZATION OF THE *ZEA MAYS* EMBRYO SAC USING FLUORESCENT PROBES AND MICROINJECTION OF LUCIFER YELLOW INTO THE FEMALE CELLS

E. Matthys-Rochon, C. Digonnet, and C. Dumas

Laboratoire de Reconnaissance Cellulaire et d'Amélioration des Plantes RCAP INRA 23879, Université de Lyon I, Bt. 741, 5ème étage, 43 Boulevard du 11 Novembre, 69622 Villeurbanne Cédex, France

INTRODUCTION

Events during double fertilization and the first stages of embryogenesis remain partly unknown because of the internal position of the female gametes in the sporophyte. The two male gametes from the pollen tube fuse with the two female gametes in the embryo sac within the ovule. Structural data from *Plumbago* sp. demonstrated that the plastid rich male gamete preferentially fuses with the egg cell leading to the embryo (Russell, 1985). In maize BMS line, the male gamete bearing the non-disjoined B-chromosome preferentially fuses with the egg cell (Roman, 1948). These results suggest a male gamete dimorphism and recognition events at the gametic level.

One way to better understand the gametic fusion processes is to isolate viable gametophytes or gametes (see review by Theunis *et al.*, 1991) and to use *in vitro* fertilization (Matthys-Rochon, 1992). The first attempt was developed by Keijzer *et al.* (1988) in *Torenia* sp.. In this work, sperm cells were released from germinated pollen by osmotic bursting. With the aid of microforceps the embryo sac was held and deposition of male gametes was achieved with microforged microcapillaries, but thus far no development of the fertilized embryo sac has been obtained. Other experiments were carried out in maize by Kranz *et al.* (1991a,b) who succeeded with *in vitro* electrofusion of gametes and production of artificial zygotes. The latter are able to divide in a nurse culture medium and, at the present time, a derived callus is developing roots. Another system to mimic double

Biotechnology Applications of Microinjection, Microscopic Imaging, and Fluorescence, Edited by P.H. Bach *et al.*, Plenum Press, New York, 1993

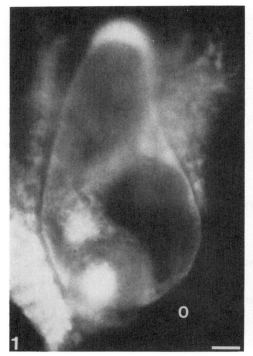

Figure 1. *Zea mays* isolated embryo sac. Positive FCR test: the cytoplasm of the cells is fluorescent. Bar: 10 μm.

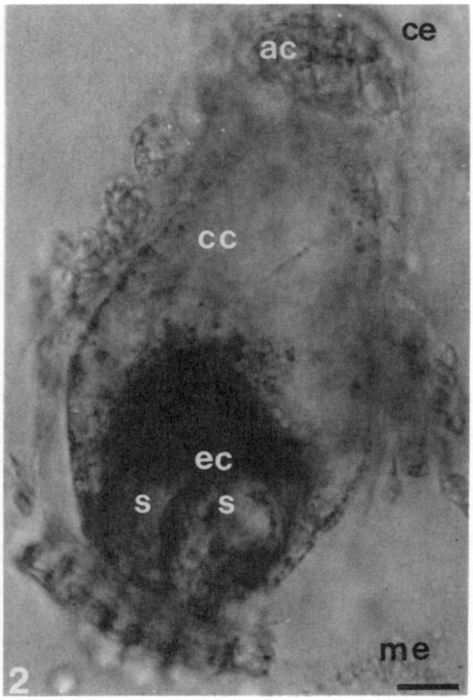

Figure 2. *Zea mays* isolated embryo sac. Phase contrast observation*. Bar: 10 μm.

In figures 2-4 please note the following: the antipodal cells (ac) at the chalazal end (ce), the central cell (cc) below and the two synergids (s) at the micropylar end (me). The egg cell (ec) is behind the synergids (s).

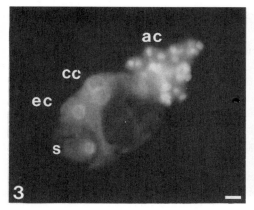

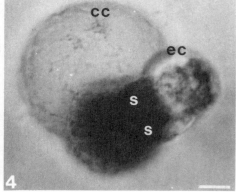

Figure 3. *Zea mays* female cells. Ethidium bromide fluorescence. Note the nuclei of the antipodal cells, central cell, egg cell and a synergid*. Bar: 10

Figure 4. *Zea mays* female cells. After enzymatic treatment and micro- dissection the female cells are released intact (phase contrast)*. Bar: 10 μm.

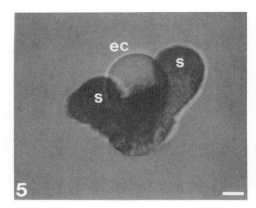

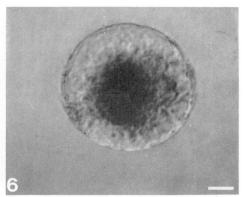

Figure 5. *Zea mays* female cells. The egg apparatus: the two synergids are linked to the egg cell (phase contrast). Bar: 10 μm.

Figure 6. *Zea mays* female cells. The egg cell (phase contrast). Bar: 10 μm.

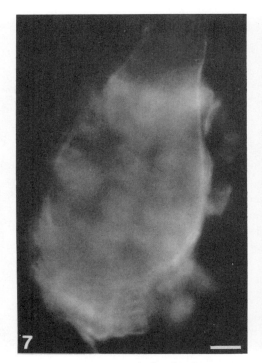

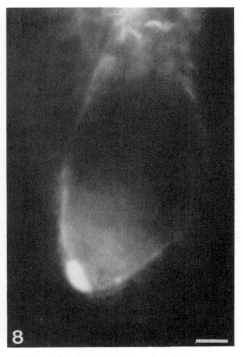

Figure 7. *Zea mays* isolated embryo sac. Calcofluor white staining: A thin layer of cellulose is visible around the embryo sac. Bar: 10 μm.

Figure 8. *Zea mays* isolated embryo sac. Aniline blue staining: A callose thin wall is visible. Note a plug at the micropylar end. Bar: 10 μm.

fertilization is to microinject nuclei into the female embryo sac cells. With that aim in mind a technical procedure in maize has been developed and is presented in this paper.

MATERIALS AND METHODS

Ear Collection

Zea mays L., A 188 hybrid line, was cultivated in growth chambers. The inflorescences were bagged prior to silk emergence to ensure an unpollinated condition. Receptivity of the ear was estimated by the silk length according to Russell (1979). Ovules were selected from the upper ears with emerged silks of 12-14 cm length.

Embryo Sac Isolation

The procedure used to isolate embryo sacs was based on Wagner *et al.* (1989). The ovaries are cut in series, row by row, and the ovules removed with a needle and incubated in a solution containing cellulase 0.05% w/v, macerozyme 0.05% w/v and pectolyase 0.025% w/v in an aqueous solution at 24°C for 15 minutes. The digested ovules are rinsed twice in BK medium (Brewbaker and Kwack, 1963) supplemented with 14% w/v sucrose (designated S14 if added to the medium) for osmotic pressure balance, and 10 mM 3-(N-morpholino)propane sulfonic acid (MOPS, Sigma Chemical Company), for pH 7 adjustment. Ovules are then microdissected with insect needles and maintained at 4°C.

Viability Test

The viability of the embryos was assessed with the FCR test (fluorochromatic reaction test) using fluorescein diacetate (Heslop-Harrison and Heslop-Harrison, 1970).

Cytological Analysis

Embryo sac nuclei were stained with 0.2% w/v aqueous ethidium bromide observed using an epifluorescence microscope (Nikon, Labophot Type 104) with a B2IF filter at 460-485 nm. Cellulose was detected by aqueous (0.001% w/v) calcofluor white solution (Nagata and Takebe, 1970). Callose deposition was stained using 0.05% (w/v) decolorized aniline blue fluorescence (Dumas and Knox, 1983) using a Carl Zeiss fluorescence inverted microscope (IM2) with an F395 excitation filter and LP 420 barrier filter.

Microinjection Procedure

In order to succeed in microinjection, embryo sacs were collected, laid on a coverslip and embedded in a 0.8% w/v low melting point agarose (Sigma Chemicals) in buffered osmotically balanced medium (BKS14 MOPS). The embryo sacs were then quickly orientated in agarose, so that the female cells can easily be reached with the microcapillary. The coverslip is then placed in a fridge and, after several minutes, the medium solidifies and the embryo sacs are immobilized, ready to be injected.

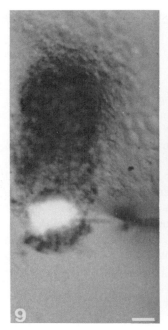 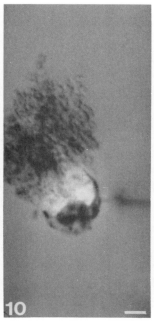 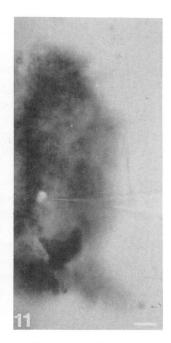

Figure 9. *Zea mays* isolated embryo sac. Lucifer Yellow micro-injection into the egg cell. Bar: 10 μm

Figure 10. *Zea mays* isolated embryo sac. Lucifer Yellow micro-injection into central cell. Bar: 10 μm.

Figure 11. *Zea mays* isolated embryo sac. Lucifer Yellow micro-injection into the central cell nucleus. Bar: 10 μm.

Microcapillaries, with a filament inside, were either pulled with a Flaming Brown micropipette puller (Sutter Instruments Co., Model P.80 PC) or purchased from Eppendorf (Femtotips). The outer diameter of the microcapillary tip was between 0.5-1 μm. All the subsequent manipulations were carried out with two micromanipulators. Microinjection was by a 5242 Eppendorf microinjector, with pressure levels adjusted to 4500 hPa for injection and 120 hPa for negative pressure. The microcapillaries were filled with a Lucifer Yellow solution (5% w/v, Lucifer Yellow CH dilithium salt, Sigma Chemicals). The fluorescent probe was injected into the embryo sac and its diffusion was followed with an epifluorescence system equipped with FTT 510 excitation filter and LP 520 barrier filter in parallel with phase contrast observation (Matthys-Rochon *et al.*, 1987).

RESULTS

Characterization of the Embryo Sac

Viability test

Using a short time enzymatic maceration, we were able to isolate viable pear-shaped mature embryo sacs in which female cells were positively stained with fluorescein diacetate and were therefore FCR positive. The cytoplasm was clearly fluorescent but not the vacuoles. Around the embryo sac, some nucellar cells (ovule tissue) may remain and also respond positively to the FCR test. Similar results were obtained after agarose embedding (Figure 1).

The Embryo Sac Cells

Phase contrast observation of isolated embryo sacs facilitates the recognition of the different female cells (Figure 2). Ethidium bromide staining of their nuclei (Figure 3) reveals the egg and the two synergids, which are uninucleate and form the egg apparatus. They are situated at the micropylar end of the embryo sac, the end through which the pollen tube enters. Surrounding the egg apparatus and occupying the centre of the embryo sac is the binucleate central cell with a large vacuole. At the far end of the embryo sac (chalazal end) are the binucleate antipodal cells, which are outside the embryo sac envelope. They are often released into the medium after enzyme treatment. As a matter of fact the integrity of the embryo sac is a function of time in the enzyme solution. If the maceration duration increases, and a manual dissection is applied, the embryo sac envelope is broken and the four contained female cells (two synergids, egg cell and central cell) may be isolated as a whole (Figure 4). The egg apparatus may also be detached, i.e. separated from the central cell as shown in Figure 5 and, finally, the egg cell may be released alone (Figure 6).

Embryo Sac Envelope

Calcofluor and aniline blue stainings show a slender fluorescent layer around the entire embryo sac (Figures 7,8) which demonstrates that the latter is enclosed in a wall-like structure composed of at least two types of glycans: cellulose and callose. In Figure 8 a dense fluorescent spot is visible (arrow) at the micropylar end and corresponds to a callose plug.

Microinjection Experiments

In order to microinject into the female cells, the embryo sacs were immobilized in a low melting point agarose medium. In that condition the microcapillary can penetrate into targeted female cells. We succeeded in microinjecting Lucifer Yellow into egg cells (Figure 9), central cells (Figure 10) and central cell nuclei (Figure 11). No diffusion of the fluorescent dye was observed either between cells or through the embryo sac envelope. As a control for further experiments, sterile water was injected into the embryo sac and the latter survived for 1-2 hours as demonstrated with the FCR test. In this type of experiment, the survival of the female gametophyte is conditioned by the environmental conditions (medium, temperature and pH) and also by the accuracy of manipulation. In that way, the microcapillary must be firmly pulled through the wall and then delicately pushed on to the cell target. The injection pressure is then activated and the fluorescent dye spreads within the central or egg cell or in the two central cell nuclei which are known to be partially fused in maize. Finally, to avoid cell injury, the microcapillary must be very gently removed.

DISCUSSION

Microinjection experiments using the very special plant material, the female gametophyte of maize, require a precise knowledge of its organization. After enzymic maceration and subsequent dissection, the embryo sac may be perfectly isolated from the ovular tissue or still surrounded with a few nucellar cells. The presence of this sporophytic tissue is not a problem for microinjection if the female cells are well orientated, i.e. if they emerge from the agarose surface. Our cytological approach shows that the embryo sac is surrounded with a wall-like structure consisting of cellulose (calcofluor white staining) and callose (aniline blue staining). At the micropylar end a callose plug is present, probably at the filiform apparatus position. The wall-like structure, which gives shape to the embryo sac, in fact constitutes a barrier to the microcapillary. In addition, TEM (transmission electron microscope) observations have demonstrated that the egg cell is limited by a partial wall, which is thickest at the micropylar end (Diboll and Larson, 1966). Consequently, in order to microinject a liquid into the egg cell, it is necessary to orientate the microcapillary in the direction of its chalazal end, and in all cases it will first have to go through the embryo sac envelope, then cross the targeted cell membrane and its thin wall.

The technical aspect we had to solve is the holding of the embryo sac. During previous assays the embryo sacs were held with a microforged pipette in a liquid medium. This system is also used for protoplasts (Crossway *et al.*, 1986). However, under these conditions, when the microcapillary was pushed through the wall, the whole embryo sac bent and microinjection was not feasible. The wall-like structure appeared to be both resistant and soft. Consequently, we searched for another technique and immobilized the embryo sacs in the semi-solid agarose medium.

With the technical procedure described above (see Material and Methods), accurate microinjections of Lucifer Yellow have been successful. According to our observations, the injected Lucifer Yellow did not seem to diffuse between and outside the cells as judged by light microscopy. To confirm this observation, it should be necessary to identify the fate of the Lucifer Yellow within the cells with the aid of electron microscopical detection (Owen, 1991). Such an investigation may allow us to determine if the dye goes through the cellular connections (plasmodesmata) which exist especially between the cells of the egg apparatus (Diboll and Larson, 1966).

In conclusion, the data presented here provides information on the structure of the maize embryo sac and shows that microinjection into the female cells is difficult, but feasible, without causing damage. Our future experiment will be to transfer, through the microinjection technique, male nuclei into the egg cell. This "*in vitro* fertilization model" may lead to a diploid product and allow analysis of zygote formation and early embryogenesis. This may be feasible particularly as an *in vitro* method of *in situ* fertilization of the embryo sac in maize is now available (Mol *et al.*, 1992). Such studies will help in gaining a better understanding of the fertilization process in angiosperms.

ACKNOWLEDGEMENTS

We gratefully acknowledge research support provided by the "Institut National de la Recherche Agronomique" (INRA) and the Germany-France PROCOPE joint program, 1991.

REFERENCES

Brewbaker J.L., and Kwack B.H. (1963) The essential role of calcium ion in pollen germination and pollen tube growth. *Am. J. Bot.* **50**: 859-865

Crossway, A., Hauptli, H., Houck, C.M., Irvine, J.M., Oakes, J.V., and Perani, L.A. (1986) Micromanipulation techniques in Plant Biotechnology. *Biotech.* **4**: 320-333

Diboll, A.G., and Larson, D.A. (1966a) An electron microscopic study of the mature megagametophyte in *Zea mays*. *Am. J. Bot.* **53**: 391-402

Dumas, C., and Knox, R.B. (1983) Callose and determination of pistil viability and incompatibility. *Theor. Appl. Genet.* **67**: 1-10

Heslop-Harrison, J., and Heslop-Harrison, Y. (1970) Evaluation of pollen viability by enzymatically-induced fluorescence: intracellular hydrolysis of fluorescein diacetate. *Stain Technol.* **45**: 115-120

Keijzer, C., Reinders, M.C., and Leferink-ten Klooster, H.B. (1988) A micromanipulation method for artificial fertilization in *Torenia, in:* "Sexual Reproduction in Higher Plants," M. Cresti, P. Gori, E. Pacini, eds., Springer Berlin, Heidelberg, New York, 119-124

Kranz, E., Bautor, J., and Lörz, H. (1991a) *In vitro* fertilization of single, isolated gametes of maize mediated by electrofusion. *Sex. Plant Reprod.* **4**: 12-18

Kranz, E., Bautor, J., and Lörz, H. (1991b) Electrofusion mediated transmission of cytoplasmic organelles through the *in vitro* fertilization process, fusion of sperm cells with synergids and central cells and cell reconstitution in maize. *Sex. Plant Reprod.* **4**: 17-21

Matthys-Rochon, E., Vergne, P., Detchepare, S., and Dumas, C. (1987) Male germ unit isolation from three tricellular pollen species: *Brassica oleracea, Zea mays* and *Triticum aestivum*. *Plant Physiol.* **83**: 464-466

Matthys-Rochon, E. (1992) *In vitro* fertilization in flowering plants, *in:* "Reproductive Biology and Plant Breeding", Y. Dattee, A. Gallais, and C. Dumas, eds., Springer Berlin, Heidelberg, New York, 197-204

Mol, R., Matthys-Rochon, E. and Dumas, C. (1992) *In vitro* culture of fertilized embryo sacs of maize: zygotes and two-celled proembryos can develop into plants. *Planta (*in press)

Nagata, T., and Takebe, I. (1970) Cell wall regeneration and cell division in isolated tobacco mesophyll protoplasts. *Planta* **92**: 301-308

Owen, T.P., Kathryn, Jr., Platt-Aloia, A., and Thomson, W.W. (1991) Ultrastructural localization of Lucifer Yellow and endocytosis in plant cells. *Protoplasma* **160**: 115-120

Roman, H. (1948) Direct fertilization in maize *Proc. Natl. Acad. Sci. USA* 34: 36-42

Russell, S.D. (1979) Fine structure of megagametophyte development in *Zea mays* Can. *J. Bot.* **57**: 1093-1111

Russell, S.D. 1985) Preferential fertilization in *Plumbago:* ultrastructural evidence for gamete level recognition in an angiosperm. *Proc. Natl. Acad. Sci. USA.* **82**: 6129-6132

Theunis, C.H., Pierson, E.S., and Cresti, M. (1991) Isolation of male and female gametes in higher plants *Sex Plant Reprod.* **4**: 145-154

Wagner, V.T., Song, Y.C., Matthys-Rochon, E., and Dumas, C. (1989) Observations on the isolated embryo sac of *Zea mays*. *Plant Science* **59**: 127-132

FLUORESCENT PROBES - WHERE DO THEY GO IN CELLS, AND WHY?

Richard W Horobin[1] and Fiza Rashid-Doubell[2]

[1]Department of Biomedical Science, Sheffield University Sheffield S10 2TN, UK
[2]Brain and Behaviour Research Group, Department of Biology, The Open University, Milton Keynes MK7 6AA, UK

WHAT ARE FLUORESCENT PROBES?

The fluorescent probes considered in this paper are either low molecular weight xenobiotics (e.g., rhodamine 123) or non-polymeric labelled metabolites (e.g., lipids), which are applied to living cells or organisms. Note that various other types of molecular probes, sometimes referred to as fluorescent probes, are excluded by this definition. Thus, macromolecular probes (e.g., fluorescent antibodies and lectins; nucleic acid probes) are not considered here. Neither are histochemical fluorochromes applied to fixed (i.e. dead) cells and tissues.

ARE FLUORESCENT PROBES MAGIC BULLETS?

A "Magic Bullet" would be completely specific, give no background, and work with all cell types and organisms. Sometimes, after reading catalogues or papers, you might be excused for thinking fluorescent probes are like this. Of course the problem is that biologists are so strongly motivated to want such reagents to exist. In routine practice, probes are selective not specific. This is often clear in the original papers describing any given probe. Patterns of selectivity displayed by probes are routinely influenced by time of exposure of cells to probe; probe concentration; solution factors such as pH and ionic strength; illumination of the specimen; type of cell; and by the presence of other probes. Moreover even useful probes often accumulate in more than one site.

Biotechnology Applications of Microinjection, Microscopic Imaging, and Fluorescence, Edited by P.H. Bach *et al.*, Plenum Press, New York, 1993

Some Bases of Selectivity

Since the widespread "Magic Bullet" concept is misleading, it is appropriate to offer some general points concerning the nature of selective uptake of probes. The "Magic Bullet" perspective emphasises the binding of a probe to some well defined and unambiguous target. Reality, however, is often more complex. Thus, the entry of a probe into a cell is also significant, except when we are concerned with cell-surface targets. Moreover, even after entry has been achieved, a probe will not be observable until a minimum amount of the compound is present. So accumulation is also a significant factor. Both entry and accumulation may involve phenomena quite distinct from those of target binding. In fact some selective probes never engage in selective binding. Consider a probe such as Nile Red. This dissolves in fat, but cannot be regarded as binding to individual lipid molecules.

CHOOSING AND UNDERSTANDING PROBES - GENERAL ISSUES

Choosing probes for marking cellular targets, and critically assessing possible positive and negative artifacts, remain challenging tasks. Even more demanding, possibly indeed still something of a mirage, is the rational design of fluorescent probes. The present authors propose that numerical structure-staining relations models (*cf* structure activity relationship), which describe uptake, accumulation and selective targeting of probes, offer practical help in these areas. Some general points relating to this proposition will now be spelt out, prior to discussion of more specific topics.

Structure-Staining Relationships

Modelling interactions of probes with cellular constituents often evokes thoughts of beautiful images of molecules and receptors on a video screen, or of space-filling physical models. However, the modelling style we have adopted is the pharmacologists' structure-activity relations approach, applied to issues arising when imaging live cells. The method used was to describe certain physico-chemical properties of probes in numerical terms; to expose cells to these probes; and look to see where the compounds localize. Correlations were then sought between the physico-chemistry and the staining outcomes. Some physico-chemical properties of the probes were described directly, and others were modelled by numerical parameters. Significant properties described directly in numerical form included the probe's electric charge (Z), its acid or base strength (pK), and its solubility. Other properties had to be modelled. Thus overall size, and the size of the aromatic system, were modelled by molecular weight and conjugated bond number (CBN), respectively. Hydrophilicity-lipophilicity was modelled by log P, namely the logarithm of the octanol-water partition coefficient. Note that for hydrophilic probes log P < 0, for lipophilic log P > 0. Some underlying technical details follow.

Cells, usually rat fibroblasts, were cultured on coverslips. Cell monolayers were exposed to solutions of fluorescent probes in a specially designed chamber. Fluorescence microscopy was then used to determine the fate of the probes. Over 160 probes, of extremely varied chemistry, have been investigated so far. Environmental factors such as time of exposure of cells to probes, probe concentration, and the illumination procedures have been varied. For each probe, the structure parameters noted above were obtained from the literature, or were calculated. For instance, pK values were obtained from standard compilations; log P values were calculated according to the method of Hansch and Leo (1979). The probes' patterns of entry and localisation within cells were then correlated with the probes' chemistry. It is these numerical correlations which constitute the "rule book" described in this paper. For full accounts of practical procedures see Horobin and Rashid (1990), and other papers cited in the References.

MECHANISMS AND MODELS FOR PROBE-CELL INTERACTIONS

Entry of Probes into Cells

Probes can enter cells in diverse ways. For instance they may be put into a cell by micro-injection or liposome fusion. Or they may enter freely following permeabilisation of

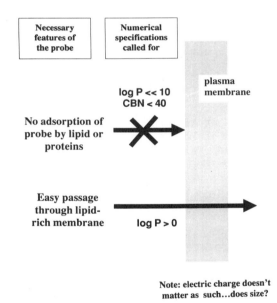

Figure 1. A simple structure-staining model specifying the features required for a fluorescent probe to enter a cell by passive diffusion through the plasma membrane.

the plasma membrane. Probes can also enter intact cells, without the aid of the experimenter. This can involve passive diffusion through the cell membrane, or be due to physiological processes. This latter may involve pores and channels; or endocytosis, either adsorptive or fluid-phase. The physico-chemical nature of those probes which can enter by passive diffusion, or by endocytosis, can be specified numerically. For a simple example, namely the case of entry by passive diffusion, see Figure 1.

Accumulation of Probes Within Cells

There are many mechanisms underlying the accumulation of probes within cells. Firstly consider processes not involving selective binding of probes to targets or receptors. Some accumulation processes involve membranes as barriers. Thus membrane encapsulation involves trapping of impermeant or colloidal species within the cell or within a membranous organelle. Here the membrane acts as a mechanical barrier. Marking of secondary lysosomes with colloidal species such as thorium dioxide or certain dyestuffs provides an example. Ion trapping also involves probes accumulating inside membrane-bound compartments. However, in this case, there must be a pH gradient across the membrane. Accumulation of weak bases in lysosomes (Rashid et al., 1991) or of weak acids within mitochondria (Rashid and Horobin, 1991) are examples of this effect. However, quite different accumulation mechanisms exist. For instance precipitation, e.g., of weak acids within low pH compartments (Rashid et al., 1991). As a final example, partitioning can occur, e.g. into the lipid of fat droplets or membranes. None of this should be taken to imply that accumulation never involves binding of probe to a target species. For instance cationic probes, such as ethidium bromide, bind intracellularly to DNA; labelled ligands can bind to cell surface receptors.

A case example. Consider the hydrophilic probes trapped within membrane bound organelles (note: how they enter the cell is another issue, requiring another numerical specification). Probes trapped in this way will have the following characteristics:

$$\log P < 0; \text{ if ionised then } pK > 7 \text{ or } pK < 7$$

As an illustration of this, consider Lucifer Yellow. This probe may be used to demonstrate fluid phase pinocytosis, since once within pinosomes this strongly hydrophilic compound ($\log P = -9.1$) does not leak. Compare this with fluorescein which, when generated from fluorescein diacetate within lysosomes, does leak, albeit slowly. Although fluorescein is hydrophilic ($\log P = -2.4$), it is a weak acid in tautomeric equilibrium with Uranil, a more lipophilic and so membrane permeant species ($\log P = +1.4$).

Selective Targeting of Probes Within Cells

Selectivity can depend on any of the factors discussed above, or on probe-target interactions, or on combinations of factors. A few examples of such mechanisms follow:

Selectivity control by binding may occur when a probe has a high affinity for a single target, with a low affinity for everything else. An example is the chelation of calcium ions by Fura salt and similar probes. An example of selectivity control by accumulation is the enzymic generation of the active form of a probe at a target site. Formation of fluorescein from fluorescein diacetate within lysosomes provides an instance. Selectivity is not due to binding of probe to the lysosome. The examples given above are relatively straightforward. However, probe selectivity can be a more complex matter. Thus a probe's pattern of selectivity may be different when confronted with an intact cell than when interacting with a cell's separated components. Such competition between targets is dramatically illustrated by the standard probes for endoplasmic reticulum, lysosomes, mitochondria, cell nuclei, and plasma membranes. All of these probes are organic cations, and in vitro they would all bind to DNA. Yet inside an intact cell the presence of a set of competing targets results in selectivity. A set of numerical specifications defining these different classes of cationic probes are given in Figure 2. These specifications are based on the work cited in the References, plus unpublished observations (Johal and Horobin).

THE RULEBOOK - ROBUST, RELIABLE AND EASY TO USE?

It is apparent that there are numerical rules describing the interactions of probes and cells. Some of the relevant work has been published (see References), some is in press. On the basis of this work, it can be stated that if the chemical structure of a fluorescent probe is

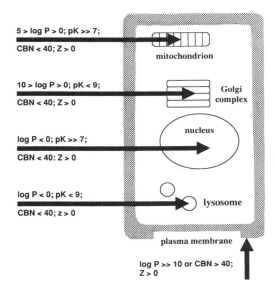

Figure 2. Numerical specifications of cationic probes selective for a range of organelles within intact cells, in terms of the probe's physico-chemical character.

known, the site(s) of the probe's localisation within a living cell may be predicted. There are corollaries of this, namely that probes can be selected rationally, that trouble-shooting the use of probes is possible and that probe design need not be so empirical. So a rulebook exists. But are the rules robust? Do they apply to all cell types? What are the limits and problems? These are questions currently being explored in several laboratories. So far the range of applicability seems wide. There are limitations of various kinds. For instance, the precision and validity of certain structure parameters is not unchallenged. This too is being explored. However, even when the rules are robust and of value, a problem remains, namely the complicated appearance of these rules. This raises the question of whether the routine benchworker will actually bother with them.

AN EXPERT SYSTEM - ONE WAY OF DEALING WITH COMPLICATIONS?

Although the rule book is complicated, it is not vastly complex. Consequently one way around the problem of complication for the routine user might be to cast the rulebook as an "advisory" or "expert" system running as a computer software package. This is currently being undertaken. A prototype version of an expert system called "Probe Advisor" has been produced, and it is now being developed and applied.

ACKNOWLEDGEMENTS

The expert system called "Probe Advisor" has been produced in collaboration with Mr Nigel Ford of the Department of Information Studies at Sheffield University.

REFERENCES

Horobin, R.W., and Rashid, F. (1990) Interactions of molecular probes with living cells and tissues. 1. Some general mechanistic proposals, making use of a simplistic Chinese box model. *Histochemistry* **94**: 205

Hansch, C., and Leo, A. (1979) "Substituent Constants for Correlation Analysis in Chemistry and Biology", Wiley, Chichester

Rashid, F., and Horobin, R.W. (1990a) Interactions of molecular probes with living cells and tissues. 2. A structure-activity analysis of mitochondrial staining by cationic probes, and a discussion of the synergistic nature of image-based and biochemical approaches. *Histochemistry* **94**: 303

Rashid, F., and Horobin, R.W. (1990b) Physiochemical features of fluorescent nuclear probes - a basis for their rational selection? *Proc. Roy. Microsc. Soc.* **26**: 58

Rashid, F., and Horobin, R.W. (1991) Accumulation of fluorescent non-cationic probes in mitochondria of cultured cells: observations, a proposed mechanism and some implications. *J. Microsc.* **163**: 233

Rashid, F., Horobin, R.W., and Williams, M.A. (1991) Predicting the behaviour and selectivity of fluorescent probes for lysosomes and related structures by means of structure-activity models. *Histochemical J.*, **23**: 450

FLUORESCENT PHOSPHOLIPIDS IN MEMBRANE

AND LIPOPROTEIN RESEARCH

A. Hermetter, R. Gorges, J. Loidl, and F. Paltauf

Department of Biochemistry and Food Chemistry
Technische Universität Graz, A-8010 Graz, Austria

INTRODUCTION

Phospholipids are the essential bilayer components of most biological membranes. In addition, they form the amphiphilic monolayer of serum lipoproteins, thus solubilizing these particles in their aqueous environment.

Figure 1 shows the chemical structures of sphingomyelin and two different glycerophospholipid subclasses carrying a choline head group. The more common diacyl glycerophospholipids contain two hydrophobic acyl chains bound to the glycerol backbone. In the alkenylacyl analogues, the acyl group in position 1 of glycerol is replaced by an alkenyl residue. Large proportions of these so-called plasmalogens are contained in the membranes of almost every animal cell (Paltauf, 1983; Hermetter, 1988). For instance, more than 70% of ethanolamine lipids in brain and 50% of choline and ethanolamine phospholipids in heart muscle are plasmalogens.

Fluorescence spectroscopy and microscopy provide useful tools for the investigation of phospholipids in their natural environment. Phospholipids can be covalently labelled with fluorescent reporter molecules. Once inserted into a membrane or lipoprotein, the behaviour of the labelled lipid analogues can be followed directly and, in particular, visualized by fluorescence microscopy.

In principle, hydrophobic side chains and polar head groups can be labelled in phospholipid molecules. However, if the labelled lipid analogue should reliably monitor the behaviour of the natural compounds, polarity and size of the fluorophor are important criteria (Hermetter, 1990).

We prepared a series of fluorescent plasmalogens labelled at their hydrophobic acyl chains in position 2. Three of which carry hydrophobic fluorescent acyl chains, namely parinaroyl (Sklar *et al.*, 1977), diphenylhexatriene propionyl (Lentz, 1989), or

Biotechnology Applications of Microinjection, Microscopic Imaging, and Fluorescence, Edited by P.H. Bach *et al.*, Plenum Press, New York, 1993

67

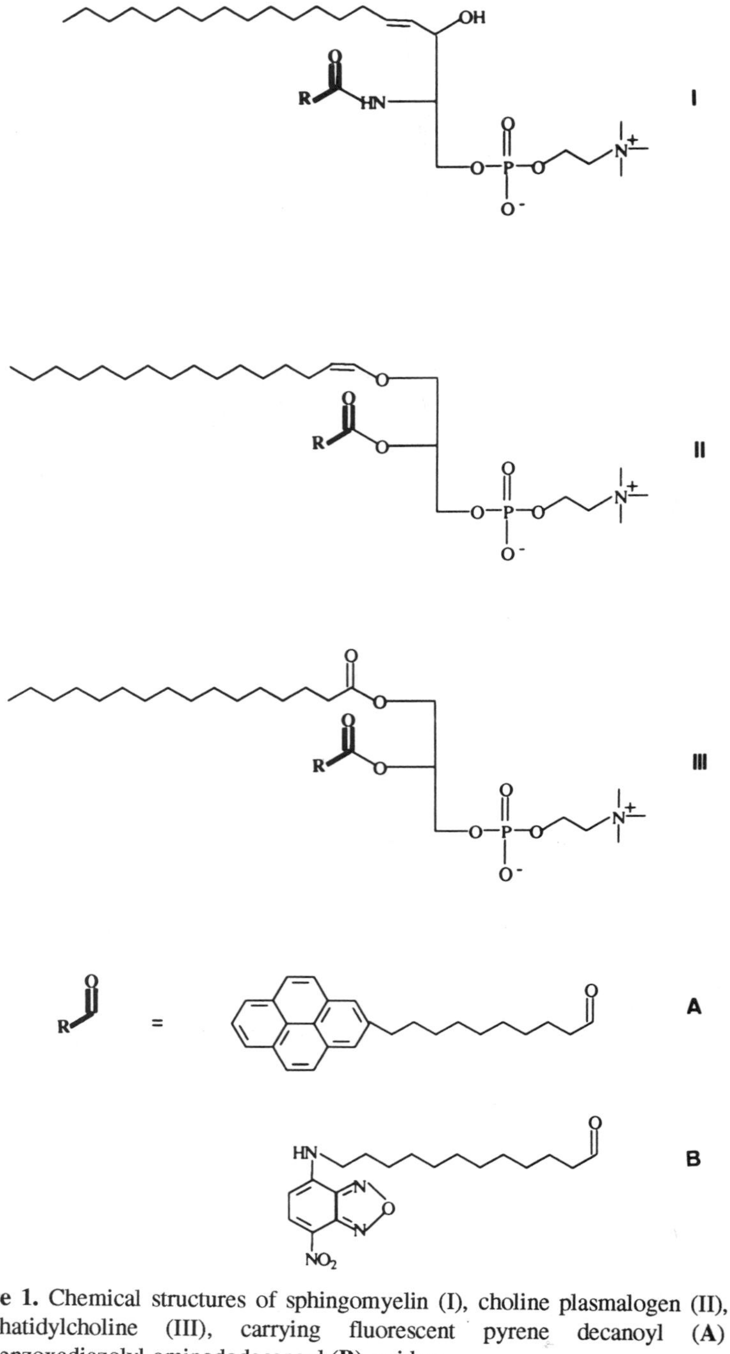

Figure 1. Chemical structures of sphingomyelin (I), choline plasmalogen (II), and phosphatidylcholine (III), carrying fluorescent pyrene decanoyl (A) or nitrobenzoxadiazolyl-aminododecanoyl (B) residues.

pyrenedecanoyl (Pownall and Smith, 1989) residues (Figure 1). However, these compounds exhibit rather unfavourable properties for fluorescence microscopy, since they absorb in the (long) UV and emit in the blue wavelength range. For microscopic observation, the yellow fluorescent 7-nitrobenz-2-oxa-1,3-diazole (NBD) derivative (Figure 1) is more appropriate (Hermetter, 1990). However, we have to take into account that the NBD residue is polar and does not warrant proper alignment of the alkyl chain by which it is linked to the glycerol moiety (Chattopadhay, 1990). It is not oriented parallel to the hydrophobic acyl chains, but folds back to the bilayer surface. This effect can be suppressed by derivatization of polar fluorophors with an additional alkyl group (Lala et al., 1988).

More recently, a series of new long wave-emitting fluorophores has been synthesized that may be of great advantage for laser scanning microscopy. Data on one of the most promising, namely 4,4-difluor-4-bora-3′,4′-diaza-s-indacene (Johnson et al., 1991), has recently been published. It is a heterocyclic boron compound showing very high quantum yields and concentration dependent emission wavelengths. In this respect, it may be used instead of pyrene, the emission spectra of which also reflect the fluorophor concentration on a molecular level. We are currently preparing phospholipids and neutral lipids labelled with this probe.

The identity and the basic properties of the parent lipid molecule are no longer maintained if the polar lipid head group is labelled with a fluorescent molecule, e.g. NBD. Although head group labelled lipids do not mimic the behaviour of their natural counterparts, they may serve as useful nonexchangeable markers (Struck and Pagano, 1980) for studying membrane-associated phenomena such as fusion or fission.

If membranes or lipoproteins are labelled with fluorescent phospholipid analogues, one has to take into account that the respective compounds form aggregates in water. Choline phospholipids form spherical multi- or single bilayers, depending on the preparation method, whereas ethanolamine lipids form flat multibilayers (Hermetter, 1990). Figure 2 shows the interaction of a vesicle containing (fluorescent) phospholipid with a lipoprotein particle, consisting of a neutral lipid core, an amphiphilic phospholipid monolayer, and specific apoproteins. Fluorescent vesicle phospholipid is incorporated into the lipoprotein surface in exchange for lipoprotein phospholipid that is transferred to the vesicle surface.

The same mechanism underlies the fluorescence labelling of artificial and biological membranes. With natural phospholipids or lipids containing hydrophobic fluorophors this process is rather slow. In general, exchange of plasmalogens is faster compared with their diacyl analogues (Szolderits et al., 1991). In contrast, phospholipids labelled with polar fluorophors, such as the NBD residue, exhibit significantly increased transfer rates.

Phospholipid exchange and therefore labelling of membranes or lipoproteins with fluorescent phospholipids is facilitated in the presence of water-soluble phospholipid transfer proteins that can be isolated from cellular cytosol or from lipoprotein-free plasma (Daum, 1990). Proteins on the surface of plasma membranes or lipoproteins may also contribute to lipid transport to some extent, since faster labelling rates are observed with the biological lipid-protein assemblies as acceptors compared with protein-free lipid vesicles.

Fluorescent Phospholipids in Cell Membranes

Two examples of the application of fluorescent phospholipids to membrane and lipoprotein studies are described here. The first deals with the role of alkenylether phospholipids in biological membranes and the second with the interaction of lipoproteins and cultured cells.

The Role of Alkenylether Phospholipids in Biological Membranes

Cultured human skin fibroblasts and animal CHO cells (Chinese hamster ovary cells) contain approximately 15 mole percent ethanolamine plasmalogen per total phospholipid. In these cells peroxisomes exhibit the full enzyme activities that are necessary for the initial steps in ether lipid biosynthesis. Cells from humans affected with peroxisomal diseases (Schutgens *et al.*, 1986), cerebro-hepato-renal Zellweger syndrome (CHRS) or rhizomelic chondrodysplasia punctata (RCDP) contain only marginal amounts (three percent) or no plasmalogen. The CHO cell mutant ZR 82 first isolated by Raetz *et al.*, are also deficient in plasmalogen (Zoeller *et al.*, 1989). If the diseased cells are supplemented with suitable ether lipid precursors, such as alkyl glycerols, they regain normal plasmalogen levels, thus providing an optimal reference system (Loidl *et al.*, 1990).

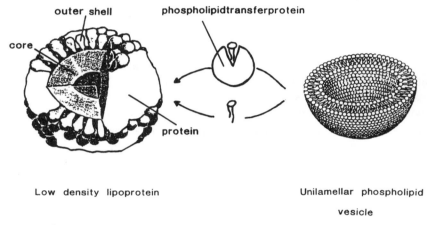

Figure 2. Spontaneous and protein-catalyzed exchange of (fluorescent) phospholipid between lipoproteins and unilamellar vesicles.

Plasmalogen-deficient cells specifically incorporate fluorescent alkenylacyl, but not diacylglycerophospholipids (Loidl *et al.*, 1990). This effect was observed with all of the labelled plasmalogens indicated above. Plasmalogen transfer is much more effective in the plasmalogen-deficient RCDP cell strain. Restoration of normal plasmalogen levels in these cells by alkylglycerol leads to a decrease in plasmalogen uptake.

Fluorescence spectroscopy can be used to show that fluorescent lipids are transferred from vesicles to the cell surface by an exchange mechanism as outlined above. The hypothesis of plasma membrane labelling could be confirmed by fluorescence microscopy, using yellow fluorescent NBD-dodecanoyl plasmalogen as a reporter system.

After very short incubation times only the plasma membrane is stained under these conditions. This effect is observed with diseased human skin fibroblasts and plasmalogen-deficient CHO cells. Thin layer chromatography analysis of a cellular lipid extract shows that the fluorescent lipid is still intact at this stage. By contrast, longer incubation times lead to staining of the entire cell interior. The internalized lipids undergo degradation followed by re-utilization of the resultant products for the synthesis of mainly triglycerides and to a minor extent phospholipids.

Phospholipids in Atherogenic Lipoproteins

Receptor-mediated endocytosis of the entire particle is a process which is crucial for the metabolism of the lipoprotein contents such as triglycerides, cholesterol esters, phospholipids, cholesterol and apoproteins (Goldstein and Brown, 1977). It is also important in the context of atherogenesis. Serum levels of atherogenic lipoproteins, in particular of low density lipoproteins, increase fatally if cells lack the specific receptor molecules which recognize and bind their apoprotein components, e.g. of the B type.

A second possible mechanism for the uptake of lipoprotein surface lipids by the cells could be an exchange process between both lipid surfaces (Owen et al., 1984). In other words cholesterol or phospholipid import might be uncoupled from the import of apoproteins and thus endocytosis of the entire particle. In order to test this hypothesis, we synthesized fluorescent NBD-dodecanoyl phosphatidylcholine and sphingomyelin (Figure 1) which are analogues of the main phospholipid classes in atherogenic lipoproteins, constituting approximately 70 and 20 percent of the surface phospholipids, respectively (Morrisett et al., 1977). The atherogenic lipoproteins, low density lipoproteins, and lipoprotein-a were labelled with these compounds, and the fate of the fluorescent lipids was studied upon interaction of the lipoproteins with cultured human skin fibroblasts.

Although low density lipoproteins and lipoprotein-a are independent risk factors for atherosclerosis, they are structurally related to each other (Utermann, 1989). Both exhibit almost identical lipid compositions and contain apoprotein B as a potential receptor binding site. Lipoprotein-a contains, in addition, apoprotein-a which is linked to apoprotein B via a disulfide bond (Sommer et al., 1991). The fate of the apoprotein component upon lipoprotein interaction with fibroblasts was followed by an immunofluorescence technique. The apoprotein B component of the lipoprotein was preloaded with anti-apo B and the resultant complex was stained with an FITC-labelled antibody against anti-apo B. After 60 minutes discrete (lipo)protein clusters can be detected on the cell surface and inside the cell.

Under the same experimental conditions where protein domains are still detected on the cell surface, the fluorescent phospholipid is homogeneously distributed over the entire plasma membrane. Therefore, an intense phospholipid flux is going on from the lipoprotein to the cell plasma membrane before the entire particle is endocytosed (Gorges et al., in press). It remains to be established whether the lipoprotein core material (neutral lipids) may also become subject to transfer by a similar mechanism.

After longer incubation times most of the lipid fluorescence is localized in lipid droplets or bound to endomembrane structures similar to endosomes or lysosomes. Endosomal compartments can be nicely visualized using Lucifer Yellow. This fluorophor is first internalized by fluid endocytosis and finally ends up in lysosomes (Stewart, 1981). From our fluorescence microscopy studies with fluorescent phospholipids we may conclude that lipoprotein surface material may enter cells by a mechanism that is different from endocytosis. Phospholipid is first transferred to the plasma membrane from where it migrates to the inner cell compartments either by fluid endocytosis or by flip-flop to the inner side of the plasma membrane, this is followed by its removal which involves cytosolic phospholipid transfer proteins.

ACKNOWLEDGEMENTS

This work was supported by a grant from the Fonds zur Förderung der Wissenschaftlichen Forschung in Österreich (project S 4615).

REFERENCES

Chattopadhay, A. (1990) Chemistry and biology of N-(7-nitrobenz-2-oxa-1,3-diazol-4-yl)-labeled lipids: fluorescent probes of biological and model membranes. *Chem. Phys. Lipids* **53**: 1-15

Daum, G., ed., (1990) Assembly of lipids into membranes. *Experientia* **46**: 551-658

Goldstein, J.L., and Brown, M.S. (1977) The low-density lipoprotein pathway and its relation to atherosclerosis. *Annu. Rev. Biochem.* **46**: 897-930

Hermetter, A. (1988) Plasmalogens. An emerging subclass of membrane phospholipids. *Comments Mol. Cell. Biophys.* **5**: 133-149

Hermetter, A. (1990) Fluorescent phospholipids. Properties and applications in membrane research. *Appl. Fluor. Technol.* **2**: 1-8

Johnson, I.D., Chang, H.C., and Haugland, R.P. (1991) Fluorescent membrane probes incorporating dipyrrometheneboron difluoride fluorophores. *Anal. Biochem.* **198**: 228-237

Lala, A.K., Dixit, R.R., Koppaka, V., and Patel, S. (1988) Design, synthesis, and fluorescence studies of fluorenyl fatty acids as new depth-dependent fluorescent probes for membranes: getting over the looping-back problem. *Biochemistry* **27**: 8981-8989

Lentz, B.R. (1989) Membrane fluidity as detected by diphenylhexatriene probes. *Chem. Phys. Lipids* **50**: 171-190

Loidl, J., Schwabe, G., Paschke, E., Paltauf, F., and Hermetter, A. (1990) Uptake of fluorescent plasmalogen analogs by cultured human skin fibroblasts deficient in plasmalogen. *Biochim. Biophys. Acta* **1049**: 75-84

Morrisett, J.D., Jackson, R.L., and Gotto, Jr., A.M. (1977) Lipid-protein interactions in the plasma lipoproteins. *Biochim. Biophys. Acta.* **472**: 93-133

Owen, J.S., McIntyre, N., and Gillett, M.P.T. (1984) Lipoproteins, cell membranes, and cellular functions. *TIBS* May: 238-242

Paltauf, F. (1983) Ether lipids in biological and model membranes, *in*: "Ether Lipids: Biochemical and Biomedical Aspects," H.K. Mangold, and F., Eds., Paltauf Academic Press, New York, 309-353

Pownall, H.J., and Smith, L.C. (1989) Pyrene-labeled lipids: versatile probes of membrane dynamics *in vitro* and in living cells. *Chem. Phys. Lipids* **50**: 191-211

Schutgens, R.B.H., Heymans,H.S.A., Wanders, R.J.A., v.d.Bosch, H., and Tager, J.M. (1986) Peroxisomal disorders: a newly recognized group of genetic diseases. *Eur. J. Pediatr.* **144**: 430-440

Sklar, L.A., Hudson, B.S., and Simoni, R.D. (1977) Conjugated polyene fatty acids: synthetic phospholipid membrane studies. *Biochemistry* **16**: 819-828

Sommer, A., Gorges, R., Kostner, G.M., Paltauf, F., and Hermetter, A. (1991) Sulfhydryl-selective fluorescence labeling of lipoprotein (a) reveals evidence for one single disulfide linkage between apoproteins (a) and B-100. *Biochemistry* **30**: 11245-11249

Stewart, W.W. (1981) Lucifer dyes-Highly fluorescent dyes for biological tracing. *Nature* **292**: 17-21

Struck, D.K., and Pagano, R.E. (1980) Insertion of fluorescent phospholipids into the plasma membrane of a mammalian cell. *J. Biol. Chem.* **258**: 5404-5410

Szolderits, G., Daum, G., Paltauf, F., and Hermetter, A. (1991) Protein-catalyzed transport of ether phospholipids. *Biochim. Biophys. Acta* **1063**: 197-202

Utermann, G. (1989) The mysteries of lipoprotein-a. *Science* **246**: 904-910

Zoeller, R.A., Allen, L.-A.H., Santos, M.J., Lazarow, P.B., Hashimoto, T., Tartakoff, A.M., and Raetz, C.R.H. (1989) Chinese hamster ovary cell mutants defective in peroxisome biogenesis. Comparison to Zellweger syndrome. *J. Biol. Chem.* **36**: 21872 -21878

SELECTION OF FLUORESCENT GOLGI COMPLEX PROBES USING STRUCTURE-ACTIVITY RELATIONSHIP MODELS

Fiza Rashid-Doubell[1] and Richard W. Horobin[2]

[1]Brain and Behaviour Research Group, Department of Biology, The Open University, Milton Keynes MK7 6AA, UK
[2]Department of Biomedical Science, The University, Western Bank, Sheffield S10 2TN, UK

INTRODUCTION

In the late 1920's certain basic supravital dyes (e.g., Neutral Red, Cresyl Blue and Methylene Blue) were observed to stain the "internal reticular apparatus", as the Golgi complex was then termed, as well as staining lysosomal contents (Parat and Painleve, 1925; Cowdry and Scott, 1928; Ludford, 1930). It was thought that such Golgi staining was due to the dyes dissolving in the lipid constituents of membranes. More recently, the fluorescent compound NBD-ceramide (i.e. N-(E-7-nitrobenz-2-oxa-1,3-di-azol-4-yl-aminocaproyl)-D-erythrosphingosine) is widely used as a Golgi complex stain (Pagano et al., 1989).

The mechanism by which this ceramide probe enters and accumulates within cells is thought to be one of lipid transfer to all other intracellular membranes, although trapping by the Golgi complex is thought to be due to the interaction with endogenous Golgi lipids (Pagano et al., 1989). However, during the course of our study, we have observed a number of compounds which also localise in the Golgi complex. The mechanisms by which they enter, accumulate and are retained will be discussed below.

Structure-activity relations models are used to provide a framework for this discussion. Certain physicochemical properties of the probes, expressed numerically, are related to the patterns of entry, accumulation and retention within cells. This approach has previously been applied to analysing the modes of actions of lysosomal (Rashid et al., 1991), mitochondrial (Rashid and Horobin, 1990; Rashid and Horobin, 1991a), and nuclear (Rashid and Horobin, 1991b) probes.

Biotechnology Applications of Microinjection, Microscopic Imaging, and Fluorescence, Edited by P.H. Bach et al., Plenum Press, New York, 1993

73

MATERIALS AND METHODS

Production of a Primary Cell Culture

A primary cell culture was derived from E18 Wistar rat embryos. Cells were dissociated using 0.25% w/v trypsin (Gibco Europe, Paisley, Scotland) and plated out at a concentration of 2.5 to 4.0 x 10^5 cells/ml into 80 ml tissue culture flasks (Nunclon Ltd, Roskilde, Denmark). Earle's growth medium was used, which consisted of 10% minimum essential medium with Earle's salts and 10% foetal calf serum (Gibco Europe).

Cells were grown in a $37^{\circ}C$ incubator with 80% humidity and 95% oxygen/5% carbon dioxide. Cells were plated out into 30 mm diameter Petri-dishes (Nunclon Ltd) containing 22 mm glass coverslips at a concentration of 0.6 to 1.0 x 10^5 cells/ml. This concentration was sufficient to allow the cells to remain sub-confluent for several days.

Obtaining Structural Parameters of Probes

Numerical structural parameters were obtained as follows. Electric charge and conjugated bond number were obtained by inspection of structures and formulae given in standard reference works (Lillie, 1977; Green, 1990). Log P values were calculated by the method of Hansch and Leo (1979). The pK values were obtained from the literature, mainly from that summarised in Perrin (1965) and Serjeant and Dempsey (1979).

Preparation of Probe Solutions

Where possible probes were made up as 0.1% w/v (typically about 2 mM) solutions in phosphate buffered saline (PBS). All solutions were filtered before being passed through a Perspex chamber containing fibroblast monolayers on coverslips at room temperature (Payne *et al.*, 1987).

Observation of Cell-Probe Interactions

The cultured rat fibroblasts were checked for autofluorescence prior to probe introduction, under all available filters. Cellular events following probe interaction were recorded with a Leitz Vario-Orthomat 2 automatic camera on Ilford HP5 film rated at 1600 ASA or Kodak Ektachrome P800/1600 also rated at 1600 ASA.

RESULTS

The 27 probes which produced staining of the Golgi complex are listed in Table 1 and micrographs illustrating typical staining is shown in Figure 1 and colour plate 2.

DISCUSSION

The twenty-seven fluorescent probes staining the Golgi complex vary greatly in their chemistry, as shown by the diverse numerical specifications given below. In the set of

Table 1 - Some physicochemical properties of probes which accumulate in the Golgi complex. Here Z is the electric charge, pK is a measure of the acid-base strength, CBN is the conjugated bond number and log P is the logarithm of the octanol-water partition coefficient.

Fluorescent Probe	Z	pK	CBN	log P
Acridine Red	+1	3.1	18	-0.63
Alizarin	0	6.2	20	+1.61
Alizarin Red S	-1	4.5; 11.0	20	-2.76
Azocarmine G	-1	---	35	-6.17
Congo Red	-2	4.1	43	-1.88
Eosin Y	-2	6.8	31	+1.4
Erythrosin A	-2	---	31	+0.66
Evans Blue	-4	---	45	-8.74
Indulin (ethanol soluble)	+1	---	51	+4.44
Metanil Yellow	-1	---	22	+0.85
Methylene Blue NN	+1	---	18	-0.72
Methylene Green	+1	3.2	21	-0.83
Neozapon Yellow GRS	0	---	23	-2.42
Neutral Red	+1	6.9	18	-0.95
Nile Blue	+1	2.4; 9.7; 10.6	23	-3.37
Nile Red	0	---	23	+5.06
Oil Red O	0	---	30	+8.99
Orcein	+1	---	36	+1.04
Phloxine	-2	---	34	+2.42
Phloxine B	-2	---	36	+3.84
Primulin	-1	---	29	+0.28
Pyronin Y	+1	---	18	-0.16
Rose Bengal	-2	4.3; 5.3	33	+4.86
Rufigallol	0	---	24	-0.76
Sirius Light Yellow RT	-2	---	59	+3.28
Sulphone Cyanin GR	-2	---	41	+0.95
Thionin (Ehrlich)	+1	6.9	18	-2.12

probes chosen for the present investigation, there appear to be five distinct classes of probe, all of which also stain other organelles.

The Five Classes of Golgi Complex Probes

Parameters are as in Table 1.

Class 1. Acidotropic weak bases, with hydrophilic cations and lipophilic free base species.
Z>0; log P $_{cation}$<0; log P $_{free\ base}$>0; CBN<40; pK$_b \approx 7$.

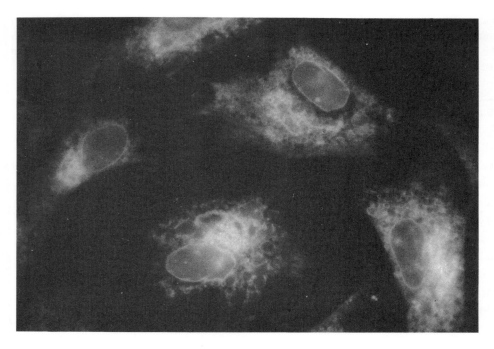

Figure 1. A fluorescence micrograph showing the staining of the Golgi complex with the acidotropic weak acid probe Phloxine. Swollen mitochondria and the nuclear envelope are also visible. Magnification x 1800.

Class 2. Acidotropic weak acids, with lipophilic least ionised species.

$Z_{nominal} \leq 0$; $+6 > \log P_{less\ ionised\ form} > 0$; CBN<40; $pK_a \approx 7$; $S < 10^{-2}$.

Class 3. Markers of adsorptive pinocytosis.

Protein binding:CBN>40.

Or Lipid binding: $\log P >> +8$.

Class 4. Markers of fluid phase pinocytosis.

$Z < 0$; $\log P < 4$;CBN< 40; amphiphilic structures only.

Class 5. Lipid partitioning probes. These are lipophilic compounds of varied physicochemistry, which do not have exceptionally high conjugated bond numbers or extremely positive log P values (ie., highly lipophilic).

Z is irrelevant; $6 > \log P > 0$; CBN< 40; non-amphiphiles only.

Mechanisms of Entry, Accumulation and Retention of Golgi Complex Probes

Weakly basic and weakly acidic probes (i.e. classes 1 and 2) enter, accumulate and are retained in the low pH portions of the Golgi complex in the same way as these probes accumulate in other low pH compartments such as lysosomes.

So the weak bases are retained by an ion-trapping mechanism (de Duve *et al.*, 1974). de Duve and co-workers suggested that these weak bases readily diffuse across biological membranes in their free base form. Then if a membrane separates two regions differing in pH, as is the case with the acidic face of the Golgi complex, the weak bases will accumulate on the acid side.

Weak acids are trapped by precipitation of insoluble free acid (Rashid *et al.*, 1991). Here the lipophilic free acid forms are membrane permeant, and entry into cells and passage through the Golgi complex membrane can occur by simple diffusion. The free acid forms of the acidotropic weak acids have low aqueous solubilities (S). Consequently, these weak acid probes are probably retained due to precipitation of their insoluble free acids in low pH compartments, a mechanism which can be termed precipitation-trapping.

Markers of adsorptive pinocytosis (class 3) first label the plasma membrane. Internalisation of membrane during endocytosis results in the labelling of secondary lysosomes and other structures. In the present study, these compounds were subsequently seen to give rise to intense fluorescence in the perinuclear region, the Golgi complex being particularly prominent. This type of membrane recycling took in the region of twenty-five minutes with the fluorescent probe ending up in the Golgi complex.

Fluid phase pinocytic markers gave a similar staining pattern to that observed with the adsorptive pinocytic markers. Presumably the mechanism by which this staining pattern was achieved is similar to the adsorptive pinocytic markers.

The fifth class of Golgi probe have lipophilicity as their shared characteristic. Hence their mode of accumulation and retention is probably by lipid partitioning, involving movement of lipophilic probes from the aqueous medium into hydrophobic sites such as lipid rich membranes. As noted, the lipid partitioning probes were chemically diverse and also stained other targets such as the cytoplasmic matrix or lysosomes.

Once the possibility of Golgi staining due to partitioning has been raised, it must be acknowledged that probes of classes 1 and 2 (ie. weak bases and weak acids) may also accumulate in the Golgi complex due to lipid partitioning, and not just by ion-trapping, or precipitation-trapping respectively. This is because the free base species of the weak bases and the free acid species of the weak acids will both be lipophilic.

Possible Applications of the Numerical Model

The selection of suitable compounds as probes, and the evaluation of staining outcomes, can now be aided by the specifications given above. In addition, the development of improved or novel Golgi complex probes should also be assisted.

ACKNOWLEDGEMENTS

We would like to thank Prof. A. Angel and Prof. S. Rose for the provision of Departmental facilities, Mr. A. Rashid for providing financial support for one of us (F.R-D.), and Rose Evison of *Change Strategies* for provision of graphics facilities. Finally, we would like to acknowledge the support of the European Commission for providing funding for this meeting.

REFERENCES

Cowdry, E.V., and Scott, G.H. (1928) Cytological studies on malaria. III. Mitochondria, Neutral Red granules and Golgi apparatus. *Ann. Inst. Pasteur de Tunis.* **17**: 233-250

de Duve, C., de Barsy, T., Poole, B., Trouet, A., Tulkens, P., and Hoof, F.V. (1974) Lysosomotropic agents. *Biochem. Pharmacol.* **23**: 2495-2531

Green, F.J. (1990) "The Sigma-Aldrich Handbook of Stains, Dyes and Indicators," Aldrich Chemical Co., Milwaukee

Hansch, C., and Leo, A. (1979) "Substituent Constants For Correlation Analysis In Chemistry And Biology," Wiley, Chichester, 18-43

Lillie, R.D. (1977) "Conn's Biological Stains," Wilkins and Wilkins Co., Baltimore

Ludford, R.J. (1930) The vital staining of normal and malignant cells. III. Vital staining of acinar cells of the pancreas and it's bearing on the the theories of vital staining with basic dyes. *Proc. Roy. Soc. (B)* **107**: 101-115

Pagano, R.E., Sepanski, M.A., and Martin, O.C. (1989) Molecular trapping of a fluorescent ceramide analogue at the Golgi apparatus of fixed cells: interaction with endogenous lipids provides trans-Golgi markers for both light and electron mircoscopy. *J. Cell Biol.* **109**: 2067-2079

Parat, M., and Painleve, J. (1925) Mise en vidence du vacuome (appareil reticulaire interne de Golgi) et du chondriome par les colorations vitales. *Bull. d'Hist. Appliq. T.* **2**: 33-47

Payne, J.N., Cooper, J.D., MacKeown, S.T., and Horobin, R.W. (1987) A temperature controlled chamber to allow observation and measurement of the uptake of fluorochromes into live cells. *J. Microsc.* **147**: 329-335

Perrin, D.D. (1965) "Dissociation Constants of Organic Bases In Aqueous Solution," Butterworths, London

Rashid, F., and Horobin, R.W. (1990) Interactions of molecular probes with living cells and tissues. 2. Structure-activity analysis of mitochondrial staining by cationic probes, and a discussion of the synergistic nature of image-based and biochemical approaches. *Histochemistry* **94**: 303-308

Rashid, F., and Horobin, R.W. (1991a) Accumulation of non-cationic probes in mitochondria of cultured cells: observations, a proposed mechanism and some implications. *J. Microsc.* **163**: 233- 241

Rashid, F., and Horobin, R.W. (1991b) Physicochemical features of fluorescent nuclear probes - a basis for their rational selection. *Proc. Roy. Microsc. Soc.* **26**: 58

Rashid, F., Horobin, R.W., and Williams, M.A. (1991) Predicting the behaviour and selectivity of fluorescent probes for lysosomes and related structures by means of structure-activity models. *Histochem. J.* **23**: 450-459

Serjeant, E.P., and Dempsey, B. (1979) "Ionisation Constants of Organic Acids In Aqueous Solution," Pergamon Press, Oxford

FLUORESCENT PROBES FOR THE ASSESSMENT
OF THE SUBCELLULAR EFFECTS OF CHEMICALS
AND DISEASE PROCESSES *IN VIVO* AND *IN VITRO*

Peter H. Bach, C. Hugh Reynolds, and Stephen Brant

Faculty of Science, University of East London
Romford Road, London E15 4LZ, England

INTRODUCTION

Recent years have seen important developments in cell biology and a greater understanding of sub-cellular processes. The availability of an ever broadening spectrum of "vital" fluorescent probes (i.e. fluorophores that do not adversely affect intact cell viablility), many of which are now designed for specific biochemical investigations, has increased our understanding of molecular processes.

The use of fluorescent probes on whole cells can be of several basic types:

a) macroscopic measurements of cell suspensions, monolayers or other preparations give limited information and are subject to interference from light-scattering and absorption of exciting or emitted light. This approach is simpler and less expensive, but heterogeneous responses are masked. This masking could be a limitation where cell-cell interactions are important or might be very useful if the overall averaged changes are significant,

b) flow cytometry quantitates the distribution of fluorescence in a sample of cells and can also be used to sort the sample into different groups. The technology is much more expensive, but is now routine for many areas of research and development,

c) the approach that is currently undergoing most development focuses on individual cells. Fluorescence microscopy has undergone rapid development with increased use of computer manipulated photon counting and ratio measurements at multiple wavelengths. Only small numbers of cells or amounts of tissue are required and quantitative information can be obtained from individual living cells as they undergo changes, often without harming the tissue. When applied *in vitro* it is especially useful in those circumstances where there are few cells in primary cultures or cells are hard to grow.

Biotechnology Applications of Microinjection, Microscopic Imaging, and Fluorescence, Edited by P.H. Bach *et al.*, Plenum Press, New York, 1993

79

This chapter will concentrate on the use of fluorescent probes for studying cellular changes brought about by chemicals or disease processes studied at the single-cell level. An overview of the use of fluorescent probes will be followed by selected examples of recently published data to illustrate the variety of applications of fluorescent probes currently being used to investigate cell pathology, pharmacology and toxicology *in vitro*.

CLASSIFICATION OF PROBES

Fluorescent probes provide a wide spectrum of useful ways of visualising both cellular structures and molecular processes.

Nomenclature

Names of probes are long and complex (to all but the organic chemist familiar with IUPAC rules), as a result of which many abbreviations or generic names are used.

Basic Fluorophores

There are a limited number of fluorophores used for a variety of purposes. These include carbocyanine, carboxyfluorescein, eosine, ethidium, fluorescein, lissamine, 7-nitrobenz-2-oxa-1,3-diazole and rhodamine, each of which has good fluorescent characteristics but may also have some weaknesses; for example, whereas fluorescein is photolabile, rhodamine is very stable. These fluorophores have been coupled to immunoglobulins, lipids, nucleic acids and other biological molecules to extended the types of specific information that can be obtained, but some of these reagents cannot be used on intact living cells. A recent addition to the list of such general fluorophores is the pyrenyloxytrisulphonic acid, Cascade Blue, which, due to its excellent fluorescent properties, has been conjugated to a whole range of dextrans, albumins, Fc-receptor binding proteins, antibodies, lectins, membrane receptor binding proteins, and biotin binding proteins, as well as biological particles and bacteria (Whitaker *et al.*, 1991a, b).

Other Fluorophores

There are a variety of other fluorophores the uses of which depend on their physicochemical properties or spectral changes associated with metabolic conversion or interaction with other molecules. Many of these are discussed in more detail below.

PROBLEMS WITH FLUORESCENT PROBES

Availability and Cost

Despite the large number of fluorescent probes that can be used for cellular biology, there are relatively few commercial sources. The increasing tendency to exploit intellectual property means that probes are becoming more expensive and available from fewer primary sources.

Purity of Probes

The question of the purity of biological stains has been addressed by the Stains Commission. There are now internationally used criteria for the assessment of the purity of biological stains, but this is not the case with fluorescent probes. Few of the current commercial suppliers provide sufficient data on purity. Unidentified contaminants in Fura-2 interfere with its use (Roe *et al.*, 1990). Many of the commonly used probes are optically active, but there are few data on the amount of each isomer present or on how each enantiomorph is handled by a cell. The recent availability of the α- and γ-isomers of fluorescein-methotrexate highlight how enantiomorphism may affect the distribution of probes (Fan *et al.*, 1991).

Stability and Safety

The commonly used probes are stable and safe (handled with the same care that applies to any chemical), but the stability and hazard of many others are not well documented, nor are specific methods for their safe disposal. There is an obligation for commercial suppliers to provide such information, but often this is covered by a bland statement covering their liability, but not facilitating handling or disposal. In the absence of safety data there is an increasing cost burden of catering for the unknown and disposal of materials thought to be hazardous. This is generally counterbalanced by the small amounts used.

Solubility and Concentration

Many of the fluorescent probes are poorly soluble in aqueous buffer, which could limit their use in studying cellular processes. The range of Fluorobora probes (Gallop *et al.*, 1982) are delivered to cells as complexes with carrier buffers that have "receptors" for the boronate group. Solubilisers of various types (e.g., dimethyl sulphoxide) can also be used, but may interfere with the cellular processes that are the focus of investigation. One of the major uncertainties in the use of these stains relates to the sensitivity of the visualising system, which often prompts use in very high concentrations, and they can cause artifacts.

Vital or Not, Consistent or Variable?

The name "vital" fluorescent probe can be misleading and the uninitiated need to be aware of artifacts that can be associated with these compounds. 4',6-diamidine-2-phenylindole (DAPSI) is widely used as a vital nuclear probe, but it stains formaldehyde-fixed material better than living cells (Bach, unpublished observation). Considerable care must be exercised in extrapolating the use of probes from one cell type to another and in interpreting their staining patterns in new cell types. For example, Yuan and Heath (1991) showed that whereas at low concentration DiOC-6(3) was an excellent mitochondrial stain for the oomycete hyphae of *Saprolegnia ferax*, higher concentrations were toxic and caused mitochondrial deformation rather than staining the endoplasmic reticulum. Similarly, 7-nitrobenz-2-oxa-1,3-diazole-hexanoic ceramide failed to stain Golgi bodies in these lower plant cells as normally described in mammalian cells.

There are few consolidated data on the adverse effects of vital probes on cells, or comparisons within a single laboratory using the same criteria and different plant and animal cells. The fluorescent cell-tracking stains PKH2-GL, dihydroethidium, and the bisbenzimide Hoechst 33342, significantly inhibited oxygen metabolism by thawed, cryopreserved bovine sperm cells compared with the unstained sperm samples, but Rhodamine 123 had no effects (Downing *et al.*, 1991). Similarly, probes that accumulate in lysosomes could also initiate enzyme leakage and alter cytosolic pH (Tapper and Sundler, 1990).

Very little is known about the biological consequences of photochemical reactions inside cells, but cells are sensitive to light. Shabalin *et al.*, (1990) have reported that even relatively modest He-Ne laser illumination (λ = 633 nm, intensity up to 6 J/cm^{-2}) can adversely affect cells. A relationship between the intensity and character of changes, the functional activity of cells, the irradiation dose and initial indices of cellular activity has been established by measuring activation of T-cell immunocompetent receptors to sheep red blood cells, stimulation of oxygen-dependent mechanisms of phagocytes revealed in the nitro blue tetrazolium test (NBT-test), and intracellular pH increase.

Metabolism of Fluorescent Probes

Many fluorescent probes are used for short periods only, but some are likely to be metabolised. For example, 2',7'-dichlorodihydrofluorescein can be rapidly oxidized to the fluorescent 2',7'-dichlorofluorescein, which can be exploited for assessing peroxidation processes. Depending on their structure it is possible that the distributions of the probe may be altered, while the spectral properties are maintained, or *vice versa*.

Biological Processes Interfering with Probes

Sulfamethoxazole hydroxylamine inhibits cellular esterase activity and limits the expression of those probes introduced into cells as esters (Leeder *et al.*, 1991). Attempts to monitor cell injury caused by reactive chemicals, such as N-acetyl-p-benzoquinoneimine, are compromised by the formation of aromatic polymers with fluorescent characteristics that interfere with calcium probes (Riley *et al.*, 1990).

Qualitative and Quantitative Data

In common with all fluorescence techniques quantitation is subject to many artifacts and interferences. Some of the more useful probes are those that undergo a spectral shift as part of the process that is being assessed, and compounds where dual excitation or emission wavelengths. Useful compounds with these characteristics include Nile Red and many of the ion probes for assessing di- and mono-valent cations and anions.

Rational Basis for Probe Use, Design and Interpretation of Data

In the past, the use of probes has progressed empirically or "lead" structures have been used as the fluorophore and, consequently, there has been relatively little understanding of the processes involved. It is only recently that any serious attempt has been made to predict the behaviour and selectivity of fluorescent probes for intracellular organelles (Rashid and Horobin, 1990, 1991; Rashid *et al.*, 1991) which is described in more detail elswhere in these proceedings [see Rashid and Horobin, and Horobin]. This simple physicochemical approach does not, however, include a consideration of several of the factors that have been highlighted above, e.g. metabolism, chirality, impurities, etc. In addition, specific binding to proteins or other cellular components, should it occur, is not predicted but data on such interactions could be very useful experimentally.

Experimental Artifacts and Limitations

Roe *et al.* (1990) have highlighted the limitations associated with the use of Fura-2 and ways to circumvent these. The same approach applies to the use of other probes. Astashkin *et al.* (1990) have recently shown that whereas cholera toxin and its β-subunit have been reported to increase cytosolic free Ca^{2+} concentration in several cell types, this now appears to be due to EDTA admixtures (present in all cholera toxin as a preservative) which "unquenches" the Quin-2 that normally binds heavy metal ions in the medium. Similarly, changes in the fluorescence intensity and reduction of the ratio inside cells for seminaphthorhodafluor-1-acetoxymethylester (Seksek *et al.*, 1991) have been attributed to the probe binding to cellular proteins.

FLUORESCENT PROBES IN TOXICOLOGY

The nature of toxicological pathology is that it seeks to assess the consequences of cellular injury in a non-specific way and generally avoids specific assay, as one parameter might not show important changes. By contrast, fluorescent probes give specific information on well defined aspects of biological structure or function. The choice of the approach depends on the purpose of the investigation, but should always be informative, simple, reproducible, quantitative, and of adequate capacity for the number of tests envisaged.

Intracellular Calcium-ion Concentration

Calcium-ion-sensitive photoproteins such as aequorin have been used to study intracellular calcium concentration (Cobbold *et al.*, 1988; Wier *et al.*, 1988; Ross, 1989; Blinks, 1989, 1990; Shimomura *et al.*, 1988, 1989, 1990; Suda and Kurihara, 1991), but will not be considered here. It is important to stress that microinjection is one of the key ways of introducing these probes into cells, which has limited their use.

There is a vast literature on calcium modulation in a wide variety of cells, caused by an ever increasing number of physiological and chemical agonists and antagonists using either Indo-1 (1-(2-amino-5-(6-carboxyindol-2-yl)phenoxy)-2-(2′amino-5′-methylphen oxy)-ethane-N,N,N′,N′-tetra-acetic acid)), Quin-2 or the more popular Fura-2. The introduction of Fluo-3, a highly fluorescent probe in the visible wavelength, to visualise Ca^{2+}-dynamics now makes it possible to use laser-scanning confocal microscopy and allows studies on elevations in cytosolic Ca^{2+} that can be resolved with a high degree of spatial resolution and interpreted quantitatively (Williams *et al.*, 1990). The use, limitations and ways of best using these and other probes have been reviewed (Roe *et al.*, 1990), as have the artifacts arising from the use of physicochemical constants, such as apparent dissociation constants measured at non-physiological temperatures for *in vitro* investigations (Shuttleworth and Thompson, 1991).

These probes can be used to measure the concentrations of ionised intracellular calcium ($[Ca^{2+}]_i$) in living cells and how this relates to normal extracellular calcium concentration ($[Ca^{2+}]_e$), based on a calibrated ratioing of dual excitation or emission (Grynkiewicz *et al.*, 1985). In normal resting cells, $[Ca^{2+}]_i$ is approximately 100 nM, but this may rise to levels approximating 2 µM within seconds following physiological stimulation, e.g. by hormones or injury. Cell damage can also give rise to elevation of Ca^{2+} either by leakage or transport into the cell or by intracellular calcium release. Alternatively, the normal Ca^{2+}-response to physiological signals may be affected in pathological states or by toxins; examples of these are shown below. A very early event when silica dust (quartz less than 5 µm in diameter) damages rat alveolar macrophages is the mobilization of intracellular calcium pools to increase $[Ca^{2+}]_i$, and this is correlated with cell damage as assessed by lactate dehydrogenase release (Chen *et al.*, 1991). Tetraphenylboron also causes calcium ion release from the sarcoplasmic reticulum (Liu and Oba, 1990).

Changes in $[Ca^{2+}]_i$ play an important role in a variety of biochemical reactions that lead to cellular proliferation and differentiation. The cornified envelope formation during keratinocyte development is associated with an increase in $[Ca^{2+}]_i$ level during calcium-induced differentiation in the short-term, but 1,25-dihydroxyvitamin D-3 only increased Ca^{2+} levels on a long-term basis (Pillai and Bikle, 1991). These data support the hypothesis that $[Ca^{2+}]_e$ regulates keratinocyte differentiation, acutely increasing their $[Ca^{2+}]_i$ levels. The effects of 1,24-dihydroxyvitamin D-3 on the synthesis of heat-shock proteins in the human monocytic line U-937, however, were not mediated by $[Ca^{2+}]_i$ (Kantengwa *et al.*, 1990).

Oxygen free radicals probably mediate ischemia-reperfusion injury to many tissues, and contribute to development of atherosclerosis, pulmonary O_2 toxicity, and a variety of reactive intermediate related chemical injury. The measurement of reactive oxygen species with redox-specific fluorescent probes is described below, but they often elevate $[Ca^{2+}]_i$. Geeraerts *et al.* (1991) have assessed the relationship between $[Ca^{2+}]_i$ and proteolysis in lethal oxidative injury in human umbilical endothelial cells during oxidative stress caused by xanthine oxidation. They showed that endothelial cell injury due to oxidative stress may be the result of Ca^{2+} influx, which could result in activation of Ca^{2+}-dependent proteases. The inhibitors leupeptin and pepstatin delayed the increase in $[Ca^{2+}]_i$ and prolonged cell viability. The volatile anaesthetics halothane, enflurane and isoflurane induce immediate, but transient, increases in $[Ca^{2+}]_i$ in rat hepatocytes (Iaizzo *et al.*, 1990).

The role of calcium has been a key to understanding neurobiology, and the actions of a number of toxins can be attributed to the release of calcium ions. Membrane depolarization has been described in the presence of caffeine (Mironov and Usachev, 1991). Ethanol-ingestion gives rise to elevated Ca^{2+} in nerve terminals (Collins and Raikoff, 1990).

A number of publications have concentrated on calcium changes in cardiac or skeletal muscle. This includes the effects of calcium channel blockers on anoxic single rat myocytes (Hano *et al.*, 1991), rat ventricular myocyte response to temperature (Liu *et al.*, 1991) the role of endothelial cells to calcium-dependent fluorescence transients in perfused rabbit hearts (Lorell *et al.*, 1990), as a way of understanding the mechanisms of hydrogen peroxide and hydroxyl free radical-induced cellular injury in cardiac myocytes (Josephson *et al.*, 1991) and the fact that oxyhaemoglobin contributes to abnormal contractions and/or irreversible damage in cultured rat aortic smooth-muscle cells (Takenaka *et al.*, 1991).

Calcium changes have also been assessed in the reproductive system. This includes the effects of arachidonate metabolites on bovine luteal cells (Alila *et al.*, 1990); porcine sperm when incubated with seminal plasma and a capacitating medium (Zhou *et al.*, 1990) and cell death in neural crest cells treated with isotretinoin and 4-oxo-isotretinoin (Davis *et al.*, 1991).

Blood cells, especially erythrocytes and platelets, are convenient systems for assessing changes in pathological conditions. Schiffl (1990) found that calcium was elevated in platelets from hypertensive haemodialysis patients. Deshmukh *et al.* (1991) showed that barbiturate treatment of rat platelets perturbs agonist-induced calcium mobilization which may be linked to their inhibitory action on protein kinase C and phosphatidylinositol 4-phosphate kinase. Such investigations are now being linked to parallel studies in a range of other second messenger and receptor mediated processes. Platelet calcium and other properties have also been used to look for, abnormalities in diabetes (Mazzanti *et al.*, 1990; Caimi *et al.*, 1991) and in manic-depressives undergoing lithium therapy (Tan *et al.*, 1990), where changes were attributed to Li^+ rather than to the illness.

pH Changes in Cells

Many of the published studies on measuring intracellular pH have concentrated on using 2′,7′-bis-(2-carboxyethyl)-5(6)-carboxyfluorescein (BCECF). More recently, dual emission long-wavelength benzo(c)xanthene pH sensors, particularly semi-naphtho-rhodafluor-1-acetoxymethylester (SNARF-1-AM), have been used. These "semi-naphtho" probes are spectrally well resolved from indicators such as Fura-2 and Indo-1 for simultaneous determination of pH and $[Ca^{2+}]$ (Whitaker *et al.*, 1991b). SNARF-1-AM is now one of the most widely used probes because, with its two strong pH-sensitive peak emissions, it requires no sequential switching of excitation filters and gives a good signal-to-noise ratio. Furthermore, fluorescent drugs including amiloride, ethyl-isopropyl-amibride, 4,4′-diisothiocyanostilbene-2,2′-disulphonic acid and cinnamate analogues can be used concomitantly with no interference (Buckler and Vaughan-Jones, 1991). The application of SNARF has been critically assessed in laser microspectrofluorometry by Seksek *et al.* (1991).

BCECF has recently been used to demonstrate that *Escherichia coli* endotoxin significantly decreased pH_i in rat hepatocytes, and caused pertubation of Ca^{2+} homeostasis (Portoles *et al.*, 1991). Alexander *et al.* (1990) showed lower sodium, proton antiport activity in vascular smooth muscle cells of Wistar-Kyoto rats than spontaneously hypertensive and Wistar rats. Gastrointestinal peptides elevated Na^+-H^+ exchange activity in cultured vascular smooth muscle cells from stroke-prone spontaneously hypertensive rats (Kobayashi *et al.*, 1990). Increased lysosomal pH (caused by methylamine, and the ionophores monensin and nigericin) was sufficient to initiate lysosomal β-N-acetyl-glucosaminidase secretion and to alter pH_i. The changes in lysosomes (preloaded with fluorescein-labelled dextran) and cytosolic pH could be dissociated by varying extracellular pH and ion composition (Tapper and Sundler, 1990).

Essig-Marcello and Van Buskirk (1990) used Neutral Red and BCECF-AM with MDCK cells to probe both lysosomes and cell integrity simultaneously using a multiwell format for cytotoxic analysis. Nedergaard *et al.* (1991) observed pH_i during acid-induced cell death of cultured neurones. Boscoboinik *et al.* (1990) showed that multidrug resistant cell lines had an intracellular pH which was higher than the parental cell line, but agents that reversed multidrug resistance did not alter pHi.

Measurement of Other Ions

Potassium (Copper *et al.*, 1990), sodium (Satoh *et al.*, 1991) and magnesium (Rutter *et al.*, 1990; Jung *et al.*, 1990) ions can now be measured, but investigations to date have been primarily into physiological mechanisms and there have been few applications to toxicology or pathology. Chemical hypoxia caused by CN^- and iodoacetate rapidly increased $[Mg^{2+}]i$, caused numerous small plasma membrane blebs and decreased ATP in cultured rat hepatocytes (Harman *et al.*, 1990).

A fascinating and unusual application of a metal-binding fluorescent probe is described by Lytton *et al.* (1991) who used a desferrioximine derivative to measure iron-scavenging.

Changes in Membranes

The application of probes to study membranes is an active and growing field. Many, perhaps most, vital probes bind to membranes, as do many toxins, and a host of pathological changes also frequently affect membranes. When a fluorophore is held more rigidly, more polarisation of the exciting light is retained in the emitted light. Fluorescence-depolarisation of membrane-located probes can thus be used to detect changes in the bilayer that affect their mobility. Muller *et al.* (1990) and Caimi *et al.* (1991) compared erythrocyte and platelet membranes in hypercholesterolaemic patients and normal controls, while Mazzanti *et al.* (1990) compared platelet membranes in diabetics and normals. Changes in membrane fluidity-rigidity were also seen in intestinal membranes from malnourished infant rabbits (Butzner *et al.*, 1990), retinol-deficient rat-liver microsomes (Kon' *et al.*, 1990), red-cell membranes after haemodialysis (Peuchant *et al.*, 1990) and in renal failure (Hagihara *et al.*, 1990), cancer cell-lines of different metastatic potential (Calderon *et al.*, 1991), bacteria differing in drug-resistance (Mukherjee *et al.*, 1989) and after irradiation of cells (Sunguov *et al.*, 1989). Alcohols and other toxins (Castro *et al.*, 1990; Kusner *et al.*, 1991; Silberman *et al.*, 1990), primaquine enantiomers (Agarwal *et al.*, 1990) and insecticides (Bondy *et al.*, 1990) caused changes in membrane fluidity. The cytoplasm, in contrast, appears to remain highly fluid and not very different from water (Fushimi and Verkman, 1991).

Transmembrane potentials can, with caution, be examined with potential sensitive probes such as the oxonol DiBAC4(3), the carbocyanine DiOC5(3), and (for mitochondrial membrane potentials) Rhodamine 123. Such methods have been used to investigate the mechanisms by which silica particles kill P388-D1 cells (Gleva *et al.*, 1990), to study the effect of hyperthermia on CHO cells (Quirt and Mackillop, 1991) and aging on mouse lymphocytes (Witkowski and Micklem, 1990). Rhodamine 123, which is generally regarded as a mitochondria-specific probe, has been used to assess mercuric chloride toxicity in MDCK cells (Lachowiez *et al.*, 1989), and together with Nonyl Acridine Orange for the determination of mitochondrial efficiency in ageing cells, during the cell cycle and in the presence of drugs (Maftah *et al.*, 1990).

Probes can readily be used to measure lipid uptake, e.g. into atherosclerotic lesions (Jaakkola and Nikkari, 1990) and by hypoxic tumour cells (Freitas *et al.*, 1990). Nile red can be used in a rapid and simple method for assessing drug-induced, lipid-rich structures in LLC-PK1 cultured cells exposed to cyclosporin-A and cremophor its vehicle (Nassberger *et al.*, 1991), and to demonstrate the presence of lipid droplets in renal medullary interstitial cells that appears to predispose them to oxidative injury (Bach *et al.*, 1986; Bach and

Kwizera, 1988; Bach, 1989, 1991; Bach and Wilks, 1992; Gregg *et al.*, 1991). Comparison of probes that bind to the membrane surface and core showed different degrees of penetration of local anaesthetics into synaptosomal membranes (Yun *et al.*, 1990). The rate of uptake of the probe DSP-6 by erythrocyte membranes was enhanced in myocardial-infarction patients, and in healthy volunteers by propranolol (Bluma *et al.*, 1990). Interactions in the liver of phenobarbital, tocopherol and lidocaine in ischaemia were studied by Sharapov *et al* (1990) by a variety of approaches including use of the probe 1,8-ANS. Riodipine, a fluorescent probe which itself possesses hypotensive activity, could identify differences between lymphocytes which were attributed primarily to their membrane content (Belevich *et al.*, 1989).

Haemolysis was used as a model system to study bile-acid cytotoxicity, where probe fluorescence changes showed that fusion of micelles with the membrane was an essential component of the toxicity mechanism (Van der Meer *et al.*, 1991).

Transport Processes

Probes may be useful for identifying malignant cell populations expressing multidrug resistance arising from a gene which encodes P-glycoprotein, an energy-dependent drug efflux pump. Kessel *et al.* (1991) showed that P388 murine leukemia could be differentiated from P388/R (an anthracycline-resistant subline) by the intensity of dye uptake (cationic rhodamine ester, cyanine, and styrylpyridinium dyes were used). Thus, many cationic and neutral fluorescent probes are substrates for the enhanced outward drug transport system. Impaired dye accumulation could be eliminated by treatment of P388/R cells with verapamil (which blocks the efflux pump). These probes could distinguish CEM cells (a human lymphoblastic leukemia line) from CEM/VLB100, a subline of CEM expressing drug resistance, but not CEM/VM-1 which lacks an enhanced efflux transporter. Similarly, the intensity of Hoechst 33258 was 40% greater in sensitive P388 cells compared to those resistant to doxorubicin (Donenko *et al.*, 1990). Rhodamine 123, daunorubicin, and doxorubicin were localized to both plasma membrane and intracellular compartments in multidrug-resistant cells. Verapamil increased the intracellular levels of probes, whereas the relative distribution of the anthracyclines between the cell surface and intracellular structures did not change. Rhodamine relocalised to intracellular sites (Weaver *et al.*, 1991).

5'-S-(2-Aminoethyl)-N-6-(4-nitrobenzyl)-5'-thioadenosine-x-2-fluorescein is a new fluorescent probe for the equilibrative inhibitor-sensitive nucleoside transporter (Wiley *et al.*, 1991), but so far only appears to have been used for flow cytometry.

Probes have been used to study the formation of pores by complement (Sauer *et al.*, 1991a,b), the mechanism of electroporation (Tekle *et al.*, 1990) and the sizes of pores induced by electroporation using FITC-labelled dextrans of different molecular weights (Rosemberg and Korenstein, 1990).

Interactions at the nucleic acid binding site of the avian retroviral nucleocapsid protein have been studied using 4,4'-bis(phenylamino)(1,1'-binaphthalene)-5,5'-disulphonic acid (Secnik *et al.*, 1990).

Redox Changes and Reactive-oxygen Species

Redox changes, oxidative stress, and the presence and consequences of reactive oxygen species is now known to be central to a range of cellular processes, the understanding of which is starting to develop with the aid of appropriate fluorescent probes. The nonfluorescent 2′,7′-dichlorodihydrofluorescein can be oxidized to the fluorochrome 2′,7′-dichlorofluorescein and has been used in a number of studies. Despite the consensus that oxidative stress is part of the injury process, Swann and Acosta (1990) failed to show intracellular reactive oxygen species (or cellular malondialdehyde) in primary cultures of renal cortical epithelial cells exposed to gentamicin. Intralobular heterogeneity of sinusoidal oxygenation and cytochrome P-450 has also been visualised in perfused rat livers with oxidative stress induced by carbon tetrachloride (Suematsu *et al.*, 1991).

Oxidation of intracellular hydroethidine by superoxide anion (O_2-) and H_2O_2 has been measured in alveolar macrophages and mononuclear phagocytes (Kobzik *et al.*, 1990). The fluorescent polyunsaturated fatty acid, parinaric acid, has been used as a peroxidation probe to measure oxidative stress in the membrane of intact erythrocytes from normal and sickle Hb-containing human erythrocytes (van den Berg *et al.*, 1991); the protective effects of vitamins E and C were also examined (van den Berg *et al.*, 1990). Oxygen-radical damage on ageing (Lebel and Bondy, 1991; Bondy and Lebel, 1991) and due to organometallic compounds (Lebel *et al.*, 1990) was studied in rat brains.

Hypoxic tumour cells are radioresistant and do not respond to oxidative anticancer treatment. Interest has centred on ways of assessing hypoxia, modulating it and using the relatively reductive nature of this tissue for therapeutic advantages. Many of the histochemical characteristics of hypoxic tumor cells have now been described by Freitas *et al.* (1990) using several methods, including Acridine Orange to visualize the morphology of an ascites and a solid tumour. Using the fluorescence intensity of nitroacridine, Katoh (1990) has measured oxygen concentration in cells as a way of estimating the radioresistancy of tumors. Hodgkiss *et al.* (1991a,b) have evaluated a range of novel fluorescent 2-nitroimidazoles to help identify hypoxic tumour cells and as the basis of chemotherapy. The compounds were rapidly taken up and concentrated in hypoxic cells where they could undergo bioreduction and bind to cellular macromolecules.

Probes for Enzyme Activity *In Situ*

β-Galactosidase is used frequently as a marker for recombinant gene expression (Zhang *et al.*, 1991) and its assay can also be applied to investigating lysosomal storage diseases (Miller *et al.*, 1991). Acridine Orange and 3-(2,4-dinitroanilino)-3′-amino-N-methyldipropylamine have been used to study the role of cathepsin D in digesting the extracellular matrix in large acidic vesicles in breast cancer cells (Montcourrier *et al.*, 1990).

Fluorescent Ligands for Receptors

Fluorescence-labelled derivatives of specific ligands are used frequently to study cell physiology, and are likely to find increasing applications in pathology. For example, the expression of D-mannose binding sites on human spermatozoa from fertile and infertile patients has been compared (Tesarik *et al.*, 1991). A range of fluorescent D_1- and D_2-dopamine ligands have been used to study the distribution of the receptors at the cellular and subcellular level, and receptor mobility in membranes in normal and diseased states (Bakthavachalam *et al.*, 1991). Fluorescent lectin probes have recently shown altered carbohydrate composition of the cochlear tectorial membrane in perinatal hypothyroidism

(Gil-Loyzaga *et al.*, 1990). Fluorescent phallacidin (a cytoskeletal probe) was used by Andre *et al.* (1990) to study cell changes on adhesion or under mechanical stress.

Immunocytochemistry: Fluorescent Antibodies

The use of an antibody provides specificity, and the coupling (directly or indirectly) of a fluorescent ligand allows detection and measurement. This vast and very important area is beyond the scope of this chapter, but it is important to point out that if living cells are being studied then normally only cell-surface antigens can be observed. Alternatively, the fluorescent antibody could be microinjected into the cell; such studies may in the future prove useful for toxicity investigations in carefully-defined situations. Practical aspects of this method have been reviewed recently by Davidson and Hilchenbach (1990).

Nucleic Acids: Fluorescence *In Situ* Hybridisation (FISH)

This technique parallels immunocytochemistry, but the reagent is a fluorescence-labelled oligonucleotide or nucleic acid that can bind specifically to complementary sequences of nucleic acids (Brandriff *et al.*, 1991; Wetmur, 1991). Chemiluminescent detection can also be used (Iwen *et al.*, 1991) to compare methods for detecting chlamydial infections in a range of cultured cells. Dual-colour investigation of cytomagalovirus DNA and mRNA (Raap *et al.*, 1991) could distinguish intron and exon sequences. FISH has also been applied to the detection and analysis of origin of diagnostic markers of tumors in human male germ carcinoma (Mukherjee *et al.*, 1991). A major advantage of the method is in rapid screening of many interphase cells, without requiring dividing cells as with conventional karyotypic analysis. Amplification of an oncogene, ERBB2, has been studied using FISH (Kallioniemi *et al.*, 1992). Numerical chromosome aberrations in fresh specimens of human solid tumours, and the ploidy profiles, were in accordance with flow cytometric DNA histograms of the tumor cells (van Dekken *et al.*, 1990).

The use of FISH in diagnosis, prognosis, and in studies of pathological tissues seems certain to grow rapidly. However, as with immunocytochemistry, its use on intact cells will require special techniques such as microinjection.

WHERE ARE PROBES GOING?

Crystal ball gazing is always limited and often wrong, but it is obviously attractive to suggest some of the areas where the use of vital fluorescent probes will be extended and the type of data that will be widely available in the not too distant future.

Most of the techniques that have been described in this chapter have been a part of the cell biologist's armamentarium for less than 10 years. Recent advances in cell biology upon which future advances will depend include:-

i) the application of patch clamping techniques to gain a greater insight into the electrophysiology of cells and membrane changes associated with chemical insult. The recent development and introduction of the patch clamp technique has added an additional aspect to the investigation of the biochemical processes that reside in individual cells. At present very little use has been made to study renal cells exposed to toxins that affect them *in vivo,*

ii) computer based image analysis of the movement and changes in shapes of cells, that will allow us to better understand the role the cytoskleton plays in normal cell function,

iii) the electrofusion of cells that will permit the combination of features unique to two different cells that are not normally present. Thus cell lines could be combined to give, for example, a transport process and a metabolic activation system to test the hypothesis that both must be present in a cell for a specific type of toxicity to occur,

iv) real time multi-optical system imaging and image manipulation to combine the most important feature of different illumination systems used for viewing: for example, to enhance the fluorescent image in terms of differential interference, contrast and clarity,

v) to combine highly sensitive and selective physicochemical methods to study the changes in metabolites that take place in the perfusion medium that is passed over cells as they are being exposed to chemical insult.

Instrumentation is becoming increasingly available to facilitate the quantitation of processes at the subcellular level. The sensitivity of photon counting imaging instrumentation in biomedical research is such that it can already provide image luminometry to allow intracellular ATP bioluminescence and chemiluminescence to be measured in parallel with $[Ca^{2+}]_i$, $[Na^+]_i$, and pH_i using living cells and time-lapse imaging to help understand the kinetics and mechanisms in response to toxins in heterogeneus cell populations (Wick, 1989; Maly *et al.*, 1989; Suematsu *et al.*, 1987). Similarly, metabolic imaging from a sandwich of frozen tissue sections covered with sections of frozen enzyme solutions and coupled to a luciferase-based light generating reaction (Mueller-Klieser *et al.*, 1988) will give the spatial distribution of glucose, lactate and ATP in normal and abberrent tissues (Wick, 1989). These exciting new developments include the use of firefly "reporter" luciferase bioluminescence to directly visualize transcription and translation in living cells (de Wet *et al.*, 1987) through the oxidation of D-luciferin in the presence of ATP (Koncz *et al.*, 1987). The chapter by White *et al.* (1992) demonstrates the potential of this technique.

The potential future uses of the single cell approach are exciting for understanding the mechanisms of selective injury and general cytotoxicity, and hence in developing rational *in vitro* tests for evaluating chemical safety and improving drug design.

CONCLUSIONS

The foregoing examples have illustrated the diversity of applications of fluorescent probes in the last few years. There is also a vast number of uses that have not been presented here, and equally the use of this tool for plant material opens a unique series of technical solutions, some of which have fundamental applications to animal cell biology, to better understand cell degenerative processes.

ACKNOWLEDGEMENTS

We are indebted to Mimps E. van Ek for typing the manuscript. The authors' research was supported by The European Commission BRIDGE and BAP programmes, The Humane Research Trust, The Dr. Hadwen Trust for Humane Research and Johns Hopkins Center for Alternatives to Animals in Testing, and in part by The Wellcome Trust, The

Smith-Kline Foundation, the University Federation for Animal Welfare and the Fund for Replacement of Animals in Medical Experiments.

REFERENCES

Agarwal, S., Khardori, R., and Agarwal, S.S. (1990) Usefulness of MC-540 fluorescent dye as probe versus scanning electron microscopy for assessing membrane changes. *Tox. Lett. (Amsterdam)* **54**: 169-174

Alexander, D., Gardner, J.P., Tomonari, H., Fine, B.P., and Aviv, A. (1990) Lower sodium, proton antiport activity in vascular smooth muscle cells of Wistar-Kyoto rats than spontaneously hypertensive and Wistar rats. *J. Hypertens.* **8**: 867-872

Alila, H.W., Corradino, R.A., and Hansel, W. (1990) Arachidonic acid and its metabolites increase cytosolic free calcium in bovine luteal cells. *Prostaglandins* **39**: 481-496

Andre, P., Capo, C., Benoliel, A.M., Buferne, M., and Bongrand, P. (1990) Analysis of the topological changes induced on cells exposed to adhesive or mechanical stimuli. *Cell Biophys.* **16**: 13-34

Astashkin, E.I., Surin, A.M., Mikhna, M.G., Nikolaeva, I.S., Lazarev, A.V., and Gukovskaya, A.S. (1990) Cholera toxin and its B subunit do not change cytosolic free calcium concentration. *Cell Calcium* **11**: 419-424

Bach, P.H. (1989) Fluorescent probes for understanding the mechanisms of chemically induced cellular injuries. *Marine Environ. Res.*, **28**, 351-356.

Bach, P.H. (1991) Better diagnosis and mechanistic understandings of nephrotoxicity using histochemistry and cytochemistry. *Prog. Histochem. Cytochem.*, **23**, 178-186.

Bach, P.H. and Kwizera, E.N. (1988) Nephrotoxicity: A rational approach to *in vitro* target cell injury in the kidney. *Xenobiotica*, **16**, 685-698.

Bach, P.H., Ketley, C.P., Dixit. M. and Ahmed, I. (1986) The mechanisms of target cell injury in nephrotoxicity. *Food Chem. Toxicol.*, **24**, 775-779.

Bach, P.H. and Wilks, M.L. (1992) *In vitro* techniques for screening for nephrotoxicity, mechanism of injury and new discovery. In: "In Vitro Toxicology: An Alternative to Animal Testing?", (Edit. Jolles, G and Cordier, A.), pp. 59-91. Academic Press, London.

Bakthavachalam, V., Baindur, N., Madras, B.K., and Neumeyer, J.L. (1991) Fluorescent probes for dopamine receptors: synthesis and characterization of fluorescein and 7-nitrobenz-2-oxa-1,3-diazol-4-yl conjugates of D-1 and D-2 receptor ligands. *J. Med. Chem.* **34**: 3235-41

Belevich, G.V., Kosnikov, V.V., Duburs, G.Y., and Dobretsov, G.E. (1989) Riodipine - a novel fluorescence probe for the identification of differences between lymphocytes. *Byull. Eksp. Biol. Med.* **108**: 597-600

Berg, J.J.M. van den, Kuypers, F.A., Lubin, B.H., Roelofsen, B., and Op-den-Kamp, J.A. (1991) Direct and continuous measurement of hydroperoxide-induced oxidative stress on the membrane of intact erythrocytes. *Free Radic. Biol. Med.* **11**: 255-261

Berg, J.J.M. van den, Kuypers, F.A., and Roelofsen, B. (1990) The cooperative action of vitamins E and C in the protection against peroxidation of parinaric acid in human erythrocyte membranes. *Chem. Phys. Lipids* **54**: 309-320

Blinks, J.R. (1989) Use of calcium-regulated photoproteins as intracellular Ca^{2+} indicators. *Methods Enzymol.* **172**: 164-203

Blinks, J.R. (1990) Use of photoproteins as intracellular calcium indicators. *Environ. Health. Perspect.* **84**: 75-81

Bluma, R.K., Kalninya, I.E., Shibaeva, T.N., and Sominskii, V.N. (1990) Use of the fluorescent probes DSP-6 for studies of the erythrocyte membrane. *Biologicheskie Membrany* **7**: 47-50

Bondy, S.C., and Lebel, C.P. (1991) Oxygen radical generation as an index of neurotoxic damage. *Biomed. Environ. Sci.* **4**: 217-23

Bondy, S.C., McKee, M., and Davoodbhoy, Y.M. (1990) Prevention of chemically induced changes in synaptosomal membrane order by ganglioside GM-1 and alpha tocopherol. *Biochim. Biophys. Acta* **1026**: 213-219

Boscoboinik, D., Gupta, R.S., and Epand, R.M. (1990) Investigation of the relationship between altered intracellular pH and multidrug resistance in mammalian cells. *Brit. J. Cancer* **61**:568-572

Brandriff, B.F., Gordon, L.A., and Trask, B.J. (1991) DNA sequence mapping by fluorescence *in situ* hybridization. *Environ. Mol. Mutagen.* **18**: 259-62

Buckler, K.J., and Vaughan-Jones, R.D. (1990) Application of a new pH-sensitive fluoroprobe (carboxy-SNARF-1) for intracellular pH measurement in small, isolated cells. *Pfluegers Arch. Eur. J. Physiol.* **417**: 234-239

Butzner, J.D., Brockway, P.D., and Meddings, J.B. (1990) Effects of malnutrition on microvillus membrane glucose transport and physical properties. *Am. J. Physiol.* **259**: 940-946

Caimi, G., Serra, A., Vaccaro, F., Lo-Presti, R., Grifo, G., D'Asaro, S., Picone, F.P., and Sarno, A. (1991) Erythrocyte calcium content and red cell membrane transverse fluidity gradient in a group of subjects with chronic renal failure. *Clin. Hemorheol.* **11**: 49-54

Calderon, R.O., Grogan, W.M., and Collins, J.M. (1991) Membrane structural dynamics of plasma membranes of living human prostatic carcinoma cells differing in metastatic potential. *Exp. Cell Res.* **196**: 192-197

Castro, V.R.O., Ashwood, E.R., Wood, S.G., and Vernon, L.P. (1990) Hemolysis of erythrocytes and fluorescence polarization changes elicited by peptide toxins, aliphatic alcohols, related glycols and benzylidene derivatives. *Biochim. Biophys. Acta* **1029**: 252-258

Chen, J., Armstrong, L.C., Liu, S.J., Gerriets, J.E., and Last, J.A. (1991) Silica increases cytosolic free calcium ion concentration of alveolar macrophages *in vitro*. *Toxicol. Appl. Pharmacol.* **111**: 211-20

Cobbold, P., Woods, N., Wainwright, J., and Cuthbertson, R. (1988) Single cell measurements in research on calcium-mobilising purinoceptors. *J. Recept. Res.* **8**: 481-91

Collins, M.A., and Raikoff, K. (1990) Elevated cytosolic calcium in cerebrocortical nerve terminals of rats during prolonged ethanol ingestion. *Life Sci.* **47**: 1221-1226

Copper, C.E., Wrigglesworth, J.M., and Nicholls, P. (1990) The mechanism of potassium movement across the liposomal membrane. *Biochem. Biophys. Res. Commun.* **173**: 1008-1012

Davidson, R.S., and Hilchenbach. M.M. (1990) The use of fluorescent probes in immunochemistry. *Photochem. Photobiol.* **52**: 431-438

Davis, W.L., Jacoby, B.H., Farmer, G.R., and Cooper, O.J. (1991) Changes in cytosolic calcium, bleb formation, and cell death in neural crest cells treated with isotretinoin and 4-oxoisotretinoin. *J. Craniofacial Gen. Dev. Biol.* **11**: 105-118

Dekken, H. van, Pizzolo, J.G., Reuter, V.E., and Melamed, M.R. (1990) Cytogenetic analysis of human solid tumors by *in situ* hybridization with a set of 12 chromosome-specific DNA probes. *Cytogenet. Cell Genet.* **54**: 103-107

Deshmukh, D.S., Kuizon, S., and Brockerhoff, H. (1991) Barbiturates inhibit intracellular Ca^{2+} rise induced by thrombin in rat platelets. *Cell Calcium* **12**: 645-654

Donenko, F.V., Borovkova, N.B., Egudina, S.V., Kabieva, A.O., Krist'Ko-Sh, I., and Moroz, L.V. (1990a) Investigation of Hoechst 33258 comparative accumulation in leukemia P388 cells sensitive and resistant to doxorubicin. *Byelleten' Eksperimental'Noi Biologii I Meditsiny* **110**: 310-312

Downing, T.W., Garner, D.L., Ericsson, S.A., and Redelman, D. (1991) Metabolic toxicity of fluorescent stains on thawed cryopreserved bovine sperm cells. *J. Histochem. Cytochem.* **39**: 485-490

Essig-Marcello, J.S., and Van-Buskirk, R.G. (1990) A double-label *in situ* cytotoxicity assay using the fluorescent probes neutral red and BCEDF-AM. *In Vitro Toxicol.* **3**: 219-228

Fan, J.G., Pope, L.E., Vitols, K.S., and Huennekens, F.M. (1991) Affinity labelling of folate transport proteins with the N-hydroxysuccinimide ester of the gamma-isomer of fluorescein-methotrexate. *Biochem. J.* **30**: 4573-4580

Freitas, I., Pontiggia, P., Barni, S., Bertone, V., Parente, M., Novarina, A., Roveta, G., Gerzeli, G., and Stoward, P. (1990) Histochemical probes for the detection of hypoxic tumor cells. *Anticancer Res.* **10**: 613-622

Fushimi, K., and Verkman, A.S. (1991) Low viscosity in the aqueous domain of cell cytoplasm measured by picosecond polarization microfluorimetry. *J. Cell. Biol.* **112**: 719-725

Gallop, P.M., Paz, M.A., and Henson, E. (1982) Boradeption: a new procedure for transferring water-insoluble agents across cell membranes. *Science* **217**: 166-169

Geeraerts, M.D., Ronveaux-Dupal, M.F., Lemasters, J.J., and Herman, B. (1991) Cytosolic free Ca^{2+} and proteolysis in lethal oxidative injury in endothelial cells. *Am. J. Physiol.* **261**: 889-896

Gil-Loyzaga, P., Bueno, A.M., Broto, J.P., and Perez, A.M. (1990) Effects of perinatal hypothyroidism in the carbohydrate composition of cochlear tectorial membrane. *Hearing. Res.* **45**: 151-156

Gleva, G.F., Goodglick, L.A., and Kane, A.B. (1990) Altered calcium homeostasis in irreversibly injured P388D-1 macrophages. *Am. J. Pathol.* **137**: 43-58

Gregg, N.J., Wilks, M.L. and Bach, P.H. (1991) The mechanistic basis of chemical- and drug-induced nephrotoxicity. *In*: "Histochemistry and Immunocytochemistry: Applications to Pharmacology and Toxicology", (Edit. Bach, P.H. and Baker J.R.J.), pp. 225-254. Chapman Hall, London.

Grynkiewicz, G., Poenie, M., and Tsien, R.Y. (1985) A new generation of calcium indicators with greatly improved fluorescence properties. *J. Biol. Chem.* **260**: 3440-3450

Hagihara, R. (1990) Lower levels of membrane lateral diffusion in erythrocytes from patients with chronic renal failure. *Med. J. Kagoshima University* **41**: 289-304

Hano, O., Silverman, H.S., Blank, P.S., Mellits, E.D., Baumgardner, R., Lakatta, E.G., and Stern, M.D. (1991) Nicardipine prevents calcium loading and "oxygen paradox" in anoxic single rat myocytes by a mechanism independent of calcium channel blockade. *Circ. Res.* **69**: 1500-1505

Harman, A.W., Nieminen, A.L., Lemasters, J.J., and Herman, B. (1990) Cytosolic free magnesium, ATP, and blebbing during chemical hypoxia in cultured rat hepatocytes. *Biochem. Biophys. Res. Commun.* **170**: 477-483

Hodgkiss, R.J., Jones, G.W., Long, A., Middleton, R.W., Parrick, J., Stratford, M.R., Wardman, P., and Wilson, G.D. (1991a) Fluorescent markers for hypoxic cells: a study of nitroaromatic compounds, with fluorescent heterocyclic side chains, that undergo bioreductive binding. *J. Med. Chem.* **34**: 2268-2274

Hodgkiss, R.J., Begg, A.C., Middleton, R.W., Parrick, J., Stratford, M.R., Wardman, P., and Wilson, G.D. (1991b) Fluorescent markers for hypoxic cells. A study of novel heterocyclic compounds that undergo bio-reductive binding. *Biochem. Pharmacol.* **41**: 533-541

Iaizzo, P.A., Olsen, R.A., Seewald, M.J., Powis, G., Stier, A., and Van-Dyke, R.A. (1990) Transient increases of intracellular calcium ion induced by volatile anesthetics in rat hepatocytes. *Cell Calcium* **11**: 515-524

Iwen, P.C., Blair, T.M.H., and Woods. G.L. (1991) Comparison of the Gen-Probe PACE 2 system, direct fluorescent-antibody, and cell culture for detecting *Chlamydia trachomatis* in cervical specimens. *Am. J. Clin. Pathol.* **95**: 578-582

Jaakkola, O., and Nikkari, T. (1990) Lipoprotein degradation and cholesterol esterification in primary cell cultures of rabbit atherosclerotic lesions. *Am. J. Pathol.* **137**: 457-466

Josephson, R.A., Silverman, H.S., Lakatta, E.G., Stern, M.D., and Zweier, J.L. (1991) Study of the mechanisms of hydrogen peroxide and hydroxyl free radical-induced cellular injury and calcium overload in cardiac myocytes. *J. Biol. Chem.* **266**: 2354-2361

Jung, D.W., Apel, L., and Brierley, G.P. (1990) Matrix free magnesium ion changes with metabolic state in isolated heart mitochondria. *Biochem.* **29**: 4121-4128

Kallioniemi, O.-P., Kallioniemi, A., Kurisu, W., Thor, A., Chen, L.-C., Smith, H.S., Waldman, F.M., Pinkel, D., and Gray, J.W. (1992) *ERBB2* amplification in breast cancer analyzed by fluorescence *in situ* hybridisation. *Proc. Natl. Acad. Sci. USA* **89**: 5321-5325

Kantengwa, S., Capooni, A.M., Bonventre, J.V., and Polla, B.S. (1990) Calcium and the heat-shock response in the human monocytic line U-937. *Am. J. Physiol.* **259**: 77-83

Katoh, T. (1990) Measurement of oxygen concentration in tumor cells by fluorescent probe (nitro compound). *Nippon Acta Radiol.* **50**: 661-668

Kessel, D., Beck, W.T., Kukuruga, D., and Schulz, V. (1991) Characterization of multidrug resistance by fluorescent dyes. *Cancer Res.* **51**: 4665-4670

Kobayashi, A., Nara, Y., Nishio, T., Mori, C., and Yamori, Y. (1990) Increased sodium ion, hydrogen ion exchange activity in cultured vascular smooth muscle cells from stroke-prone spontaneously hypertensive rats. *J. Hypertens.* **8**: 153-158

Kobzik, L., Godleski, J.J., and Brain, J.D. (1990) Oxidative metabolism in the alveolar macrophage: Analysis by flow cytometry. *J. Leukocyte Biol.* **47**: 295-303

Kon', I.YA., Sokolov, A.I., Filatov, I.YU., and Deev, A.I. (1990) Vitamin A and microsomal membranes: Effects of retinol deficiency on lipid microviscosity and phospholipid turnover in rat liver microsomes. *Biokhimiya* **55**: 982-987

Koncz, C., Olsson, O., Langridge, W., Schell, J., and Szalay, A.A. (1987) Expression and assembly of functional bacterial luciferase in plants. *Proc. Natl. Acad. Sci. USA.* **84**: 131-135

Lachowiez, R.M., Clayton, B., Thallman, K., Dix, J.A., and Van-Buskirk, R.G. (1989) Rhodamine 123 as a probe of *in vitro* toxicity of MDCK cells. *Cytotech.* **2**: 203-212

Lebel, C.P., Ali, S.F., McKee, M., and Bondy, S.C. (1990) Organometal-induced increases in oxygen reactive species: The potential of 2′4′-dichlorofluorescein diacetate as an index of neurotoxic damage. *Toxicol. Appl. Pharmacol.* **104**: 17-21

Lebel, C.P., and Bondy, S.C. (1991) Persistent protein damage despite reduced oxygen radical formation in the aging rat brain. *Int. J. Dev. Neurosci.* **9**: 139-146

Leeder, J.S., Dosch, H.M., and Spielberg, S.P. (1991) Cellular toxicity of sulfamethoxazole reactive metabolites: I. Inhibition of intracellular esterase activity prior to cell death. *Biochem. Pharmacol.* **41**: 567-574

Liu, G., and Oba, T. (1990) Effects of tetraphenylboron-induced increase in inner surface charge on calcium ion release from sarcoplasmic reticulum. *Jpn. J. Physiol.* **40**: 723-736

Liu, B., Wang, L.C.H., and Belke, D.D. (1991) Effect of low temperature on the cytosolic free calcium in rat ventricular myocytes. *Cell Calcium* **12**: 11-18

Lorell, B.H., Apstein, C.S., Cunningham, M.J., Schoen, F.J. Weinberg, E.O., Peeters, G.A., and Barry, W.H. (1990) Contribution of endothelial cells to calcium-dependent fluorescence transients in rabbit hearts loaded with indo 1. *Circ. Res.* **67**: 415-425

Lytton, S.D., Cabantchik, Z.I., Libman, J., and Shanzer, A. (1991) Reversed siderophores as antimalarial agents. II. Selective scavanging of Fe (III) from parasitized erythrocytes by a fluorescent derivative of desferal. *Mol. Pharmacol.* **40**: 584-590

Maftah, A., Petit, J.M., Leprat, P., Ratinaud, M.H., and Julien, R. (1990) A new methodology for testing chemicals and drugs on cell activity. *Int. J. Cosmetic Sci.* **12**: 253-264

Maly, F.E., Vittoz, M., Urwyler, A., Koshikawa, K., Schleinkofer, L., and De Weck, A.L. (1989) A dual microtiter plate (192 sample) luminometer employing computer-aided single photon imaging applicable to cellular luminescence and luminescence immunoassay. *J. Immunol. Methods* **122**: 91-96

Mazzanti, L., Rabini, R.A., Faloia, E., Fumelli, P., Bertoli, E., and De-Pirro, R. (1990) Altered cellular calcium ion and sodium ion transport in diabetes mellitus. *Diabetes* **39**: 850-854

Meer, R. van der, Termont, D.S.M.L., and De-Vries, H.T. (1991) Differential effects of calcium ions and calcium phosphate on cytotoxicity of bile acids. *Am. J. Physiol.* **260**: G142-G147

Miller, S.P.F., French, S.A., and Kaneski, C.R. (1991) Synthesis and characterization of a novel lysosomotropic enzyme substrate that fluoresces at intracellular pH. *J. Org. Chem.* **56**: 30-34

Mironov, S.L., and Usachev, J.M. (1991) Caffeine affects calcium uptake and calcium release from intracellular stores: Fura-2 measurements in isolated snail neurons. *Neurosci. Lett.* **123**: 200-202

Montcourrier, P., Mangeat, P.H., Salazar, G., Morisset, M., Sahuquet, A., and Rochefort, H. (1990) Cathepsin D in breast cancer cells can digest extracellular matrix in large acidic vesicles. *Cancer Res.* **50**: 6045-6054

Mueller-Klieser, W., Walenta, S., Paschen, W., Kallinowski, F., and Vaupel, P. (1988) Metabolic imaging in microregions of tumors and normal tissues with bioluminescence and photon counting. *J. Natl. Cancer Inst.* **80**: 842-848

Mukherjee, K., Chakrabarty, A.N., and Dastidar, S.G. (1989) Experimental studies on membrane permeability defects of drug-resistant bacteria. *Indian J. Med. Res. Sect. A.* **89**: 238-242

Mukherjee, A.B., Murty, V.V., Rodriguez, E., Reuter, V.E., Bosl, G.J., and Chaganti, R.S. (1991) Detection and analysis of origin of i(12p), a diagnostic marker of human male germ cell tumors, by fluorescence *in situ* hybridization. *Genes Chromosom. Cancer* **3**: 300-307

Muller, S., Ziegler, O., Donner, M., Drouin, P., and Stoltz, J.F. (1990) Rheological properties and membrane fluidity of red blood cells and platelets in primary hyperlipoproteinemia. *Atherosclerosis* **83**: 231-238

Nassberger, L., Bergstrand, A., and DePierre, J.W. (1991) An electron and fluorescence microscopic study of LLC-PK1 cells, a kidney epithelial cell line: normal morphology and cyclosporin A- and cremophor-induced alterations. *Int. J. Exp. Pathol.* **72**: 365-378

Nedergaard, M., Goldman, S.A., Desai, S., and Pulsinelli, W.A. (1991) Acid-induced death in neurons and glia. *J. Neurosci.* **11**: 2489-2497

Peuchant, E., Motta, C., Salles, C., and Clerc, M. (1990) A compensatory mechanism improving red cell membrane fluidity in hemodialyzed patients. *Clin. Chim. Acta* **190**: 57-66

Pillai, S., and Bikle, D.D. (1991) Role of intracellular-free calcium in the cornified envelope formation of keratinocytes: Differences in the mode of action of extracellular calcium and 1,25 dihydroxyvitamin D-3. *J. Cell. Physiol.* **146:** 94-100

Portoles, M.T., Ainaga, M.J., Municio, A.M., and Pagani, R. (1991) Intracellular calcium and pH alterations induced by *Escherichia coli* endotoxin in rat hepatocytes. *Biochim. Biophys. Acta* **1092:** 1-6

Quirt, C.F., and Mackillop, W.J. (1991) The effect of hyperthermia on transmembrane potential Chinese hamster ovary cells *in vitro. Rad. Res.* **126:** 96-103

Raap, A.K., Van-de-Rijke, F.M., Dirks, R.W., Sol, C.J., Boom, R., and van-der-Ploeg, M. (1991) Bicolor fluorescence *in situ* hybridization to intron and exon mRNA sequences. *Exp. Cell. Res.* **197:** 319-322

Rashid, F., and Horobin, R.W. (1990) Interaction of molecular probes with living cells and tissues: Part 2. A structure-activity analysis of mitochondrial staining by cationic probes, and a discussion of the synergistic nature of image-based and biochemical approaches. *Histochem.* **94:** 303-308

Rashid, F., and Horobin, R.W. (1991) Accumulation of fluorescent non-cationic probes in mitochondria of cultured cells: observations, a proposed mechanism, and some implications. *J. Microsc.* **163:** 233-241

Rashid, F., Horobin, R.W., and Williams, M.A. (1991) Predicting the behaviour and selectivity of fluorescent probes for lysosomes and related structures by means of structure-activity models. *Histochem. J.* **23:** 450-459 check ???

Riley, R.J., Leeder, J.S., Dosch, H.M., and Spielberg, S.P. (1990) Interactions between N-acetyl-p-benzoquinoneimine and fluorescent calcium probes: Implications for mechanistic toxicology. *Anal. Biochem.* **191:** 253-261

Roe, M.W., Lemasters, J.J., and Herman, B. (1990) Assessment of Fura-2 for measurements of cytosolic free calcium. *Cell Calcium* **11:** 63-74

Rosemberg, Y., and Korenstein, R. (1990) Electroporation of the photosynthetic membrane: A study by intrinsic and external optical probes. *Biophysical J.* **58:** 823-832

Ross, W.N. (1989) Changes in intracellular calcium during neuron activity. Annu. Rev. Physiol. 51: 491-506

Rutter, G.A., Osbaldeston, N.J., McCormack, J.G., and Denton. R.M. (1990) Measurement of matrix free magnesium ion concentration in rat heart mitochondria by using entrapped fluorescent probes. *Biochem. J.* **271:** 627-634

Satoh, H., Hayashi, H., Noda, N., Terada, H., Kobayashi, A., Yamashita, Y., Kawai, T., Hirano, M., and Yamazaki, N. (1991) Quantification of intracellular free sodium ions by using a new fluorescent indicator, sodium-binding benzofuran isophthalate in guinea pig myocytes. *Biochem. Biophys. Res. Commun.* **175:** 611-616

Sauer, H., Pratsch, L., Fritzsch, G., Bhakdi, S., and Peters, R. (1991a) Complement pore genesis observed in erythrocyte membranes by fluorescence microscopic single-channel recording. *Biochem. J.* **276:** 395-399

Sauer. H., Pratsch, L., and Peters, R. (1991b) A microassay for the pore-forming activity of complement, perforin, and other cytolytic proteins based on confocal laser scanning microscopy. *Anal. Biochem.* **194:** 418-24

Schiffl, H. (1990) Correlation of blood pressure in end-stage renal disease with platelet cytosolic free-calcium concentration. *Klinische Wochenschrift* **68:** 718-722

Secnik, J., Wang, Q., Chang, C.M., and Jentoft, J.E. (1990) Interactions at the nucleic acid binding site of the avian retroviral nucleocapsid protein: Studies utilizing the fluorescent probe 4,4'-bis(phenylamino)(1,1'-binaphthalene)-5,5'-disulfonic acid. *Biochem.* **29:** 7991-7997

Seksek, O., Henry-Toulme, N., Sureau, F., and Bolard, J. (1991) SNARF-1 as an intracellular pH indicator in laser microspectro-fluorometry: A critical assessment. *Anal. Biochem.* **193:** 49-54

Shabalin, V.N., Ivanenko, T.V., Skokova, T.V., and Ol'Shanskii, A.Y. (1990) Immunological and physicochemical effects of lasers on biological objects. *Immunologiya:* 30-32

Sharapov, V.I., Grek, O.R., and Zykov, A.A. (1990) Effect of α-tocopherol and lidocaine on structural and functional alterations in liver endoplasmic reticulum membranes after induction with phenobarbital during the postischemic period. *Voprosy Meditsinskoi Khimii* **36:** 14-18

Shimomura, O., Musicki, B., and Kishi, Y. (1988) Semi-synthetic aequorin. An improved tool for the measurement of calcium ion concentration. *Biochem. J.* **251:** 405-410

Shimomura, O., Musicki, B., and Kishi, Y. (1989) Semi-synthetic aequorins with improved sensitivity to Ca^{2+} ions. *Biochem. J.* **261:** 913-20

Shimomura, O., Inouye, S., Musicki, B., and Kishi, Y. (1990) Recombinant aequorin and recombinant semi-synthetic aequorins. Cellular Ca2+ ion indicators. *Biochem. J.* **270**: 309-312

Shuttleworth, T.J., and Thompson, J.L. (1991) Effect of temperature on receptor-activated changes in intracellular calcium ion concentration and their determination using fluorescent probes. *J. Biol. Chem.* **266**: 1410-1414

Silberman, S., McGarvey, T.W., Comrie, E., and Perskey, B. (1990) The influence of ethanol on cell membrane fluidity, migration, and invasion of murine melanoma cells. *Exp. Cell Res.* **189**: 64-68

Suda, N., and Kurihara, S. (1991) Intracellular calcium signals measured with Fura-2 and aequorin in frog skeletal muscle fibers. *Jpn. J. Physiol.* **41**: 277-295

Suematsu, M., Kato, S., Ishii, H., Asako, H., Yanagisawa, T., Suzuki, H., Oshio, C., and Tsuchiya, M. (1991) Intralobular heterogeneity of carbon tetrachloride-induced oxidative stress in perfused rat liver visualized by digital imaging fluorescence microscopy. *Lab. Invest.* **64**: 167-173

Sunguov, A.Yu., Mumanzhinov, V.V., Sinakevich, I.G., and Zhuikov, A.G. (1989) Biophysical studies of surface of irradiated cells of Ehrlich ascites carcinoma. *Studia Biophys.* **129**: 5-12

Swann, J.D., and Acosta, D. (1990) Failure of gentamicin to elevate cellular malondialdehyde content or increase generation of intracellular reactive oxygen species in primary cultures of renal cortical epithelial cells. *Biochem. Pharmacol.* **40**: 1523-1526

Takenaka, K., Yamada, H., SAkai, N., Ando, T., Nakashima, T., Nishimura, Y., Okano, Y., and Nozawa, Y. (1991) Cytosolic calcium changes in cultured rat aortic smooth-muscle cells induced by oxyhemoglobin. *J. Neurosurg.* **74**: 620-624

Tan, C.H., Javors, M.A., Seleshi, E., Lowrimore, P.A., and Bowden, C.L. (1990) Effects of lithium on platelet ionic intracellular calcium concentration in patients with bipolar (manic-depressive) disorder and healthy controls. *Life Sci.* **46**: 1175-1180

Tapper, H., and Sundler, R. (1990) Role of lysosomal and cytosolic pH in the regulation of macrophage lysosomal enzyme secretion. *Biochem. J.* **272**: 407-414

Tekle. E., Astumian, R.D., and Chock, P.B. (1990) Electro-permeabilization of cell membranes: Effect of the resting membrane potential. *Biochem. Biophys. Res. Commun.* **172**: 282-287

Tesarik, J., Mendoza, C., and Carreras, A. (1991) Expression of D-mannose binding sites on human spermatozoa: comparison of fertile donors and infertile patients. *Fertil. Steril.* **56**: 113-118

Weaver, J.L., Pine, P.S., Aszalos, A., Schoenlein, P.V., Currier, S.J., Padmanabhan, R., and Gottesman, M.M. (1991) Laser scanning and confocal microscopy of daunorubicin, doxorubicin, and rhodamine 123 in multidrug-resistant cells. *Exp. Cell Res.* **196**: 323-329

Wet, J.R. de, Wood, K.V., Deluca, M., Helinski, D.R., and Subramani, S. (1987) Firefly luciferase gene: Structure and expression in mammalian cells. *Mol. Cell Biol.* **7**: 725-737

Wetmur, J.G. (1991) DNA probes: applications of the principles of nucleic acid hybridization. *Crit. Rev. Biochem. Mol. Biol.* **26**: 227-259

Whitaker, J.E., Haugland, R.P., Moore, P.L., Hewitt, P.C., Reese, M., and Haugland, R.P. (1991a) Cascade Blue derivatives: Water soluble, reactive, blue emission dyes evaluated as fluorescent labels and tracers. *Anal. Biochem.* **198**: 119-130

Whitaker, J.E., Haugland, R.P., and Prendergast, F.G. (1991b) Spectral and photophysical studies on benzo(c)xanthene dyes: Dual emission pH sensors. *Anal. Biochem.* **194**: 330-344

Wick, R.A. (1989) Photon counting imaging: applications in biomedical research. *BioTechniques*, **7**, 262-269.

Wier, W.G., Beuckelmann, D.J., and Barcenas-Ruiz, L. (1988) [Ca2+]i in single isolated cardiac cells: a review of recent results obtained with digital imaging microscopy and fura-2. *Can. J. Physiol. Pharmacol.* **66**: 1224-1231

Wiley, J.S., Brocklebank, A.M., Snook, M.B., Jamieson, G.P., Sawyer, W.H., Craik, J.D., Cass, C.E., Robins, M.J., McAdam, D.P., and Paterson, A.R.P. (1991) A new fluorescent probe for the equilibrative inhibitor-sensitive nucleoside transporter: 5'-S-(2-Aminoethyl)-N-6-(4-nitrobenzyl)-5'-thioadenosine (SAENTA)-x-2-fluorescein. *Biochem. J.* **273**: 667-672

Williams, D.A., Cody, S.H., Gehring, C.A., Parish, R.W., and Harris, P.J. (1990) Confocal imaging of ionized calcium in living plant cells. *Cell Calcium* **11**: 291-298

Witkowski, J.M., and Micklem, H.S. (1990) Transmembrane electrical potential of lymphocytes in aging mice: Flow cytometric studies with oxonol, cyanine, and rhodamine 123 dyes. *Aging Immunol. Infect. Disease* **2**: 287-294

Yuan, S., and Heath, I.B. (1991) A comparison of fluorescent membrane probes in hyphal tips of *Saprolegnia ferax. Exp. Mycol.* **15**: 103-115

Yun, I., Jung, I.K., Kim, I.S., Yum, S.M., Baek, S.Y., and Kang, J.S. (1990) The penetration of local anesthetics into the synaptosomal plasma membrane vesicles isolated from bovine brain. *Pacific J. Pharmacol.* **5**: 19-26

Zhang, Y.Z., Naleway, J.J., Larison, K.D., Huang, Z.J., and Haugland, R.P. (1991) Detecting LacZ gene expression in living cells with new lipophilic, fluorogenic beta-galactosidase substrates. *Fed. Am. Soc. Exp. Biol. J.* **5**: 3108-3113

Zhou, R., Shi, B., Chou, K.C.K., Oswalt, M.D., and Haug, A. (1990) Changes in intracellular calcium of porcine sperm during *in vitro* incubation with seminal plasma and a capacitating medium. *Biochem. Biophys. Res. Commun.* **172**: 47-53

WHOLE-CELL PATCH-CLAMPING:
INTRODUCING SUBSTANCES INTO CELLS
DURING ELECTRICAL MEASUREMENTS FROM
THE CELL MEMBRANE -A REVIEW OF
POTENTIAL DIFFICULTIES IN PLANT AND
ANIMAL CELLS

Bert Van Duijn[1,2], Can Ince[3], Zheng Wang[2], A. Freek Weidema[2], Kees R. Libbenga[1], and Dirk L. Ypey[2]

[1]Institute of Molecular Plant Sciences, Department of Botany, Nonnensteeg 3, 2311 VJ Leiden, The Netherlands

[2]Address for correspondence: Department of Physiology and Physiological Physics, Leiden University, P.O. Box 9604, NL-2300 RC Leiden, The Netherlands

[3]Department of General Surgery, Erasmus University, Dr. Molewaterplein 40, NL-3015 GD Rotterdam, The Netherlands

INTRODUCTION

The purpose of this article is to summarize the possibilities and limitations of patch electrodes for the introduction of substances into cells while measuring electrical membrane properties in, for example, the study of hormone signal transduction pathways. An introduction to the patch-clamp technique will be given. The advantages of the patch-clamp technique for electrophysiological studies are reviewed extensively in many different journals; excellent overviews are given by Neher and Sakmann (1992), and Neher (1992). This chapter will emphasize some limitations and problems which should be considered before deciding to apply the patch-clamp technique to find the answer to a specific question. Some examples of the use of the patch-clamp technique based on our experience with electrophysiological studies in animal and plant cells are presented to illustrate different measurement possibilities.

Biotechnology Applications of Microinjection, Microscopic Imaging, and Fluorescence, Edited by P.H. Bach *et al.*, Plenum Press, New York, 1993

The introduction of the patch-clamp technique by Neher and Sakmann (Hamill *et al.*, 1981; Neher and Sakmann, 1976) greatly stimulated membrane electrophysiological research on small cells and opened the way to single channel measurements. Besides its application to excitable cells (i.e. nerve and muscle cells), investigations into the electrical properties of non-excitable cells have made significant progress. The patch-clamp technique permits a "non-invasive" method of measuring ionic currents flowing through single ionic channels (Hamill *et al.*, 1981). Furthermore, measurements of membrane current or potential in the whole-cell mode are relatively free from leakage currents (Hamill *et al.*, 1981) and allow control of the cytosolic composition (e.g. ion concentrations). Use of the patch-clamp technique has revealed that ionic channels play a role in a diversity of cellular processes (e.g. Neher and Sakmann, 1992; Dunne and Petersen, 1991; Blatt, 1991). In addition to the electrical regulation of ionic channels, increasing attention is being directed to biochemical regulation by ligands, G-proteins, phosphorylation etc. of ion channel activity (e.g. Fairley-Grenot and Assmann, 1991; Ordway *et al.*, 1991; Birnbaumer *et al.*, 1991). Genetic cloning, analysis and manipulation of ionic channel proteins have also made great progress (e.g. Catterall, 1988; Unwin, 1989; Stühmer, 1991; McCormack *et al.*, 1991; Jan and Jan, 1992).

In the whole-cell configuration of the patch-clamp technique free access to the cytoplasm is obtained. This does not only allow the experimenter to measure the electrical properties of the whole-cell plasma membrane, but also provides the possibility of controlling the cytoplasmic composition and to introduce substances into the cell. In principle, these operations can be done while measuring ion channel activity or membrane potential. An experiment is possible in which, after perfusion of a cell with a calcium or pH dependent fluorescent probe, calcium or pH imaging are combined with electrical measurements (eg. Iijima *et al.*, 1990).

INTRODUCTION TO PATCH-CLAMPING

The patch-clamp technique distinguishes itself from the intracellular microelectrode technique in that a glass microelectrode (patch pipette) with a tip diameter of about 1-2 μm is placed upon the cell membrane instead of piercing through the cell membrane (Figure 1). The patch-clamp technique is based on an electrical isolation of a small membrane patch from the rest of the cell membrane. To achieve this, the patch pipette is pressed against the cell membrane, and a slight under-pressure is applied to the pipette interior. Under appropriate conditions a tight seal between the lipids of the cell membrane and the glass of the patch pipette develops. This seal, referred to as a "giga-seal" (due to its resistance being in the giga-ohm range) causes a very high (>1 Gohm) resistance between the inside of the patch pipette and the surrounding bath solution (Figure 1). The giga-seal is essential in patch-clamp measurements. The electric currents through the piece of membrane, the patch, under the patch pipette will mainly flow into the pipette when the "seal resistance" between cell and pipette is very high (i.e. when a giga-seal is present).

Furthermore, a high-seal resistance reduces the current noise of the recording (Neher, 1992). The "non-invasive" mode of the patch-clamp technique is referred to as the cell-attached patch configuration. In this configuration, currents through single ion channels in the patch can be measured in the intact cell. The Ag-AgCl electrodes in the patch pipette and the bath solution are connected to a pico-ampere resolution current-to-voltage converter. This amplifier can measure current and holds the potential at a chosen value (voltage-clamp measurements). The opening and closing of single ionic channels in the

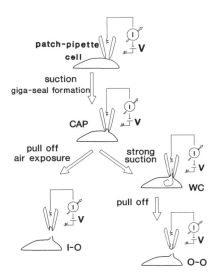

Figure 1. Schematic representation of the configurations of the patch-clamp technique. CAP = cell-attached-patch, WC = whole-cell, I-O = inside-out, O-O = outside-out.

patch will cause quantal changes in the measured current, reflecting the all-or-nothing change of the ion channel conductances (Figure 2).

Starting from the cell-attached patch configuration, another configuration with free access to the cell interior can be obtained, the whole-cell configuration. Application of extra underpressure in the cell-attached-patch configuration will break the patch. The cell is perfused by the patch pipette contents and a whole-cell configuration is achieved (Figure 1). In the whole-cell configuration the current through the whole-cell membrane can be measured with a voltage-clamp amplifier (Figure 3).

Besides current measurements membrane potential measurements can be made with a current-clamp amplifier in the whole-cell configuration. Due to the free access to the cytoplasm the whole-cell configuration is often used to introduce substances into the cytoplasm to study their effects on the plasma membrane properties (Figure 4). The rate of diffusion of the pipette contents into the cell depends on molecular size and the size of the pipette opening (Pusch and Neher, 1988).

A third configuration makes it possible to control the physical environment of the channels more precisely. This configuration is achieved by pulling the patch pipette from the cell membrane in the cell-attached-patch configuration. This leads to the inside-out excised-patch configuration. In this configuration the inside of the cell membrane is exposed to the bath saline solution. Pulling of the pipette in the whole-cell configuration results in the formation of the outside-out excised-patch configuration (Figure 1). The excised-patch configurations allows the researcher to study the ionic channels in the cell membrane independently from the processes occurring in the cell. For more detailed technical descriptions of the patch-clamp technique see Hamill *et al.* (1981), Sakmann and Neher (1983), and Standen *et al.* (1987).

SOME DIFFICULTIES WITH PATCH-CLAMPING ON PLANT AND ANIMAL CELLS

Giga-seal Formation

The formation of the giga-seal is a crucial step in patch-clamp experiments. Without a large resistance between the patch pipette and the cell membrane, patch-clamp recordings are not possible. The giga-seal formation is based on a molecular interaction between the lipids in the cell membrane and the glass of the patch pipette (Corey and Stevens, 1983). A good interaction between the membrane and the glass, leading to giga-seal formation, can

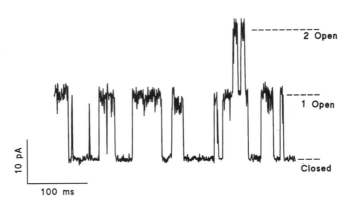

Figure 2. Single-channel currents recorded from a freshly-isolated chick osteoclast (cf. Ravesloot *et al.*, 1989 for the isolation procedure) using the cell-attached configuration of the patch-clamp technique. Outward currents were recorded with 150 mM K^+ in the patch-pipette and an extracellular-like solution in the bath at a pipette potential of -160mV. The current was filtered at 1 kHz and sampled at 2 kHZ. Up to two channels were simultaneously open in this patch as indicated by the dashed lines (1 Open, 2 Open) at the right of the trace. These channels belong to the large conductance (160 pS) Ca^{2+}-dependent K^+ channels (Weidema *et al.*, 1992).

be disturbed by several factors. Some of the factors preventing giga-seal formation have a technical nature and can be avoided. For example, the glass used for patch pipettes should be clean and the solutions should be filtered. Other problems with the formation of giga-seals are more structural and depend on the cell type used. These problems cannot always be solved. Cells which have to be maintained in organs or tissues in order to elucidate the role of ion channels in larger cell structures cannot easily be studied with the patch-clamp technique. However, recently some progress has been made with the application of the patch-clamp technique in tissue slices (Edwards *et al.*, 1989; Konnerth, 1990). Problems arise with cells which are organised in a tissue and are not easily accessible because of cell coupling or the presence of connective tissue. These cell types can be made available for the patch-clamp technique by dispersion of the tissue (Trube, 1983) or by removal of covering cells by (local) enzyme treatment (Edwards *et al.*, 1989). Dissociation

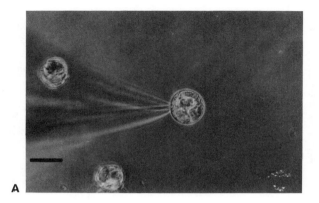

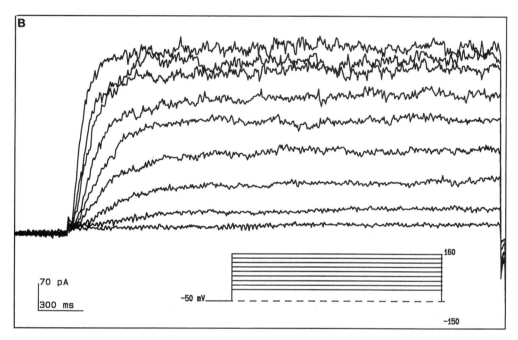

Figure 3. Patch-clamp experiment on tobacco protoplast in the whole-cell configuration. **(A)** Tobacco protoplast isolated as described before (Van Duijn *et al.*, 1992 a,b) in the whole-cell configuration. Bar indicates 25 μm. (**B**) Voltage and time dependent whole-cell outward K^+ currents (Van Duijn *et al.*, 1992 a,b). The current responses upon stepwise changes in the potential across the membrane (as indicated in the potential protocol below) are shown. The bath contained a low K^+ concentration solution and the pipette a high K^+ concentration solution. Displayed currents are corrected for the leak-conductance.

of cells from tissue can be achieved by mechanical distortion or by enzymatic treatment. The obtained single cells can be used directly for patch-clamp measurements or after some time of culturing. It is obvious that such dissociation may affect ionic channel properties. Mechanical distortion damages the cells and enzymatic treatment will also damage the cell

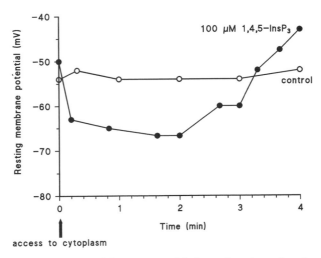

Figure 4. Membrane potential response of 1-day cultured rat dorsal root ganglion neurons upon perfusion with an intracellular-like solution (open symbols) and an intracellular-like solution with 100 μM inositol-1,4,5-trisphosphate (1,4,5-InsP3) (closed symbols) in the whole-cell configuration of the patch-clamp technique. This membrane potential hyperpolarization upon perfusion with a 1,4,5-InsP3 containing solution is likely due to the capacity of 1,4,5-InsP3 to release, even in the whole-cell configuration, calcium from intracellular stores, which activates calcium-dependent potassium channels (Wang *et al.*, 1992).

membrane and the proteins in it. Therefore, measurements on freshly isolated cells may have expressed other membrane electrophysiological properties, of the cells due to the isolation procedure. On the other hand, cells which are cultured for some time may recover from the isolation procedure, but will have additional altered physiological properties due to the culturing. It has been shown in some preparations, such as macrophages, that the expression of channels is dependent on the time cells have been put on glass (Ypey and Clapham, 1984; Gotti *et al.*, 1987; Lewis and Cahalan, 1988). Hence, cells obtained from tissues by dissociation may suffer from altered electrophysiological properties. Therefore, for a complete interpretation of the results obtained from such cells a comparison with results from undamaged and uncultured cells remains necessary in many cases.

Cells which have been shown to be difficult to patch include mostly non-animal cells, such as cellular slime molds, plant and yeast cells. Many of these cell types have an extracellular matrix of glycoproteins, collagen or other substances. Plant cells have a cellulose cell wall. Due to this the cell membrane is not reachable for the patch pipette. For measurements the cell wall of the plant cells has to be removed enzymatically to produce protoplasts which are suited for patch-clamp experiments. However, different protocols for preparation of protoplasts may give very different seal-success ratios with, in many cases, no giga-seals at all. Elzenga *et al.* (1991) describe a successful method for preparing protoplasts suitable for patch-clamp measurements. In addition, the ion concentrations *in vivo* at the plasma membrane may be different from the ion concentrations in the bath solution, due to the ion exchange properties of the cell wall and the surface charges of the plasma membrane (Hedrich *et al.*, 1987; Sentenac and Grignon, 1981). Therefore, care must be taken with the interpretation of patch-clamp measurements on protoplasts in relation to the *in vivo* situation. Among eukaryotes, yeast cells and cellular slime mold cells (*Dictyostelium* sp.) can be relatively easily manipulated genetically as compared with other eukaryote cell types. Patch-clamp measurements on these cells, therefore, may greatly advance the knowledge of channel structure, function, and regulation when combined with molecular genetic techniques available for these cells. Application of the patch-clamp technique to these cells is, however, difficult since the cell wall of yeast cells has to be removed enzymatically (Gustin *et al.*, 1986). Giga-seal formation on cellular slime molds is very difficult, probably due to the presence of glycoproteins on the cell surface and strong cell movement. The formation of giga-seals is only possible on cells bathed in unphysiological saline solutions [such as extreme high calcium concentrations of >10 mM which have been described Müller *et al.* (1986), Van Duijn (1990), Yoshida and Inouye, (personal communication)]. Problems with seal-formation can also arise with contractile cells (heart cells, muscle cells) and mobile cells. Muscle cells have to be treated with drugs (e.g. tetrodotoxin) to prevent muscle contraction during patch-clamp measurements (Hamill *et al.*, 1981). Amoeboid cells which move relatively fast, e.g. *Dictyostelium discoideum*, "walk away" from the patch pipette. These problems are cell type dependent and need alteration of the physiological status of the cells to be solved. Hence application of the patch-clamp technique in these cells will require control experiments to test the presence of normal physiological and functional properties (e.g. gene expression, metabolism, hormone responses) under the same conditions.

In summary, electrical measurements from cells on which giga-seal formation can only be achieved under non-physiological conditions such as enzyme treatment, cell culture and different ionic conditions should be interpreted with care.

Influence of the Patch Pipette on the Cell Properties

The presence of the patch pipette on the cell surface might change the electrophysiological properties of the cell membrane as well as evoke other reactions of the cell. An influence on channel kinetics due to the presence of the patch pipette for acetylcholine channels in *Xenopus* myocytes (Rojas and Zuazaga, 1988) and complete inactivation of potassium channels in rat pituitary pars intermedia cells (Cota and Armstrong, 1988) has been reported. These effects might also be present for other channels and other cell types. Plasma membrane hyperpolarizations in plant cells and protoplasts subjected to mechanical pressure have been demonstrated (Zimmermann and Steudle, 1974; Pantoja and Willmer, 1986). The existence of the patch pipette on the cell membrane can cause a "crawling away" reaction of the cell. In addition, enhanced membrane movement, which can only be observed by time-lapse photography, or crimpling up of cells can be present. During the formation of the giga-seal, stretch sensitive ionic channels are activated in macrophages (Ince *et al.*, 1988) and tobacco protoplasts (unpublished observations). In other cell types stretch-activated ionic channels are also reported (Medina and Bregestovski, 1988; Morris and Sigurdson, 1989; Taglietti and Toselli, 1988). Therefore, some time has to elapse after giga-seal formation to achieve a new steady-state situation. These reactions of cells on the presence of the patch pipette, described above, will disturb the measurements, not only due to a possible giga-seal loss, but even more due to changed physiological properties of the cell. Hence, the behaviour of the cells during cell-attached-patch measurements should always be observed (e.g. with time-lapse video) carefully, to exclude artifacts.

Achievement of Whole-cell Configuration and Cell Perfusion

To measure whole-cell membrane currents or potential in the whole-cell configuration, which provides free access to the cytoplasm, the membrane patch under the patch electrode has to be broken (e.g. by negative pressure or a high electric voltage pulse). In some cases this membrane disruption can be difficult (e.g. when a high extracellular calcium concentration is required as in *Dictyostelium* and some higher plant cells). The success rate of making the whole-cell configuration can vary from day-to-day in other cell types as well. In the ectoderm of *Xenopus* blastula, whole-cell recording is not possible due to clogging of vitelline granules in the electrode (Baud, 1987). Furthermore, resealing of the broken membrane can occur.

After breakage of the membrane patch, perfusion of the cell with the patch electrode solution starts. Diffusion time constants depend on molecular size, cell size and size of the pipette opening (Oliva *et al.*, 1988; Pusch and Neher, 1988). The perfusion of the cell with the patch-pipette content has clearly many advantages and opens new ways to investigate the involvement of electrical processes in cell functioning. Because of the perfusion, the ionic composition of the electrode solution should be quite similar to that of the cytoplasm. However, for most cell types this composition is still unknown, and can vary between different cells within a species and between the same cell types in different species (Ince *et al.*, 1987). Among changes in the ionic composition, disturbance of the intracellular calcium concentration (Ca_i) and pH (pH_i) and clamping of Ca_i (by using EGTA) and pH_i (by using HEPES buffer) are most serious. This may especially be a problem when the effects of hormones or drugs on plasma membrane electrical properties are studied, since changes in Ca_i and pH_i likely play an important role in transmembrane signal transduction and cellular functioning. In addition, effects of HEPES buffer on chloride channels have been reported (Yamamoto and Suzuki, 1987), which shows that the pipette solution must be chosen with care. Not only will contents of the patch pipette diffuse into the cell, cytoplasmic constituents will also leave the cell and diffuse into the pipette. A real steady-state situation, therefore, will never be achieved in the whole-cell configuration, and loss of unknown intracellular components may impair essential cellular processes and/or the process under study (Fenwick *et al.*, 1982; Lindau and Fernandez, 1986; Penner *et al.*, 1987). Loss of large intracellular components (proteins and even organelles) can be prevented by use of the so called "slow whole-cell", in which the membrane patch is not disrupted but permeabilized with e.g. nystatin or ATP (Lindau and Fernandez, 1986; Buisman *et al.*, 1988; Horn and Marty, 1988). However, in this configuration substances cannot be introduced into the cytoplasm.

CONCLUSION

The patch-clamp technique in the whole-cell configuration is a very powerful way to study electrical membrane properties and regulation of these membrane properties by different intracellular substances and during different cell physiological processes. The perfusion of the cell with the pipette solution is the major advantage of measurements in the whole-cell configuration, which can be used to introduce substances into cells during electrical membrane measurements. However, this perfusion can also be a disadvantage because of the loss of cytoplasmic components (e.g. macromolecules, small organelles) due to diffusion into the patch-pipette. Undesirable buffering of ion concentrations (potential second messengers, such as Ca^{2+}, H^+) may be a problem too. Technical limitations in the application of the patch-clamp technique are mostly due to difficulties in giga-seal formation and cell damaging side-effects of procedures to make cells accessible to the patch-pipette (e.g. removal of the cell wall of plant cells, or dissociation of cells in tissues).

ACKNOWLEDGEMENTS

The authors wish to thank Dr. M. Wang for critical reading of the manuscript. BVD and KRL are partly supported by EEC Bridge contract BIOT CT90-0158C. BVD was financially supported by NWO through BION/SVB THEMA I (Intracellular conduction of signals) project 811-416-112.

REFERENCES

Baud, C. (1987) Ionic basis of membrane potential in developing ectoderm of the *Xenopus* blastula. *J. Physiol.* **393**: 525-544

Birnbaumer, L., Abramowitz, J., and Brown, A.M. (1990) Receptor-effector coupling by G proteins. *Biochim. Biophys. Acta* **1031**: 163-224

Blatt, M.R. (1991) Ion channel gating in plants: physiological implications and integration for stomatal function. *J. Membrane Biol.* **124**: 95-112

Buisman, H.P., Steinberg, T.H., Fischbarg, J., Silverstein, S.C., Vogelzang, S.A., Ince, C., Ypey, D.L., and Leijh, P.C.J. (1988) Extracellular ATP induces a large non-selective conductance in macrophage plasma membranes. *Proc. Natl. Acad. Sci. (USA)* **85**: 7988-7992

Catterall, W.A. (1988) Genetic analysis of ion channels in vertebrates. *Annu. Rev. Physiol.* **50**: 395-406

Corey, D.P., and Stevens, C.F. (1983) Science and technology of patch-recording electrodes, *in:* "Single channel recording," B. Sakmann and E. Neher, eds., Plenum Press, New York, 53-68

Cota, G., and Armstrong, C.M. (1988) Potassium channel "inactivation" induced by soft-glass patch pipettes. *Biophys. J.* **53**: 107-109

Dunne, M.J., and Petersen, O.H. (1991) Potassium selective ion channels in insulin-secreting cells: physiology, pharmacology and their role in stimulus-secretion coupling. *Biochim. Biophys. Acta* **1071**: 67-82

Edwards, F.A., Konnerth, A., Sakmann, B., and Takahashi, T. (1989) A thin slice preparation for patch clamp recordings from neurons of the mammalian central nervous system. *Europ. J. Physiol.* **414**: 600-612

Elzenga, J.T.M., Keller, C.P., and Van Volkenburgh, E. (1991) Patch clamping protoplasts from vascular plants. *Plant Physiol.* **97**: 1573-1575

Fairley-Grenot, K., and Assmann, S.M. (1991) Evidence for G-protein regulation of inward K^+ channel current in guard cells of *Faba* bean. *The Plant Cell* **3**: 1037-1044

Fenwick, E.M., Marty, A., and Neher, E. (1982) Sodium and calcium channels in bovine chromaffin cells. *J. Physiol.* **331**: 599-635

Gotti, C., Sher, E., Cabrini, D., Bondiolotti, G., Wanke, E., Mancinelli, E., and Clementi, F. (1987) Cholinergic receptors, ion channels, neurotransmitter synthesis, and neurite outgrowth are independently regulated during the *in vitro* differentiation of a human neuroblastoma cell line. *Differentiation* **34**: 144-155

Gustin, M.C., Martinac, B., Saimi, Y, Culbertson, M.R., and Kung, C. (1986) Ion channels in yeast. *Science* **233**: 1195-1197

Hamill, O.P., Marty, A., Neher, E., Sakmann, B., and Sigworth, F.J. (1981) Improved patch clamp techniques for high resolution current recording from cells and cell free membrane patches. *Europ. J. Physiol.* **391**: 85-100

Hedrich, R., Schroeder, J.I., and Fernandez, J.M. (1987) Patch-clamp studies on higher plant cells: a perspective. *TIBS* **12**: 49-52

Horn, R., and Marty, A. (1988) Muscarinic activation of ionic currents measured by a new whole-cell recording method. *J. Gen. Physiol.* **92**: 145-159

Iijima, T., Sand, O., Sekiguchi, T., and Matsumoto, G. (1990) Simultaneous recordings of cytosolic Ca^{2+} level and membrane potential and current during the response to thyroliberin in clonal rat anterior pituitary cells. *Acta Physiol. Scand.* **140**: 269-278

Ince, C., Thio, B., Van Duijn, B., Van Dissel, J.T., Ypey, D.L., and Leijh, P.C.J. (1987) Intracellular K^+, Na^+, Cl^- concentrations and membrane potential in human monocytes. *Biochim. Biophys. Acta* **905**: 195-204

Ince, C., Coremans, J.M.C.C., Ypey, D.L., Leijh, P.C.J., Verveen, A.A., and Van Furth, R. (1988) Phagocytosis by human macrophages is accompanied by changes in ionic channel currents. *J. Cell Biol.* **106**: 1873-1878

Jan, L.Y., and Jan, Y.N. (1992) Structural elements involved in specific K^+ channel functions. *Annu. Rev. Physiol.* **54**: 537-555

Konnerth, A. (1990) Patch-clamping in slices of mammalian CNS. *TINS* **13**: 321-323

Lewis, R.S., and Cahalan, M.D. (1988) Subset-specific expression of potassium channels in developing murine T lymphocytes. *Science* **239**: 771-775

Lindau, M., and Fernandez, J.M. (1986) IgE-mediated degranulation of mast cells does not require opening of ion channels. *Nature* **319**: 150-153

McCormack, K., Tanouye, M.A., Iverson, L.E., Lin, J-W, Ramaswami, M., McCormack, T., Campanelli, J.T., Mathew, M.K., and Rudy, B. (1991) A role for hydrophobic residues in voltage-dependent gating of shaker K^+ channels. *Proc. Natl. Acad. Sci. (USA)* **88**: 2931-2935

Medina, I.R., and Bregestovski, P.D. (1988) Stretch-activated ion channels modulate the resting membrane potential during early embryogenesis. *Proc. Roy. Soc. B.* **235**: 95-102

Morris, E., and Sigurdson, W.J. (1989) Stretch-inactivated ion channels coexist with stretch-activated ion channels. *Science* **243**: 807-809

Müller, U., Malchow, D., and Hartung, K. (1986) Single ion channels in the slime mold *Dictyostelium discoideum*. *Biochim. Biophys. Acta* **857**: 287-290

Neher, E. (1992) Ion channels for communication between and within cells. *EMBO J.* **11**: 1673-1679

Neher, E., and Sakmann, B. (1976) Single channel currents recorded from membrane of denervated frog muscle fibres. *Nature* **260**:799-802

Neher, E., and Sakmann, B. (1992) The patch-clamp technique. *Sci. Am.* **266**: 28-35

Oliva, C., Cohen, I.S., and Mathias, R.T. (1988) Calculation of time constants for intracellular diffusion in whole-cell patch clamp configuration. *Biophys. J.* **54**: 791-799

Ordway, R.W, Singer, J.J., and Walsh jr., J.V. (1991) Direct regulation of ion channels by fatty acids. *TINS* **14**: 96-100

Pantoja, O., and Willmer, C.M. (1986) Pressure effects on membrane potentials of mesophyll protoplasts and epidermal cell protoplasts of *Commelia communis* L. *J. Exp. Bot.* **37**: 315-320

Penner, R., Pusch, M., and Neher, E. (1987) Washout phenomena in dialysed mast cells allow discrimination of different steps in stimulus secretion coupling. *Biosci. Rep.* **7**: 313-321

Pusch, M., and Neher, E. (1988) Rates of diffusional exchange between small cells and measuring patch pipette. *Europ. J. Physiol.* **411**: 204-211

Ravesloot J.H., Ypey D.L., Vrijheid-Lammers T., and Nijweide P.J. (1989) Voltage activated K^+ conductances in freshly isolated embryonic chick osteoclasts. *Proc. Natl. Acad. Sci. (USA)* **86**: 6821-6825

Rojas, L., and Zuazaga, C. (1988) Influence of the patch pipette glass on single acetylcholine channels recorded from *Xenopus* myocytes. *Neurosci. Lett.* **88**: 39-44

Sakmann, B., and Neher, E., eds. (1983) "Single channel recording", Plenum Press, New York,1-503

Sentenac, H., and Grignon, C. (1981) A model for predicting ionic equilibrium concentrations in cell walls. *Plant Physiol.* **68**: 415-419

Standen, N.B., Gray, P.T.A., and Whitaker, M.J., eds. (1987) "Microelectrode techniques", The Comp. Biologists Ltd., Cambridge, U.K., 1-256

Stühmer, W. (1991) Structure-function studies of voltage-gated ion channels. *Annu. Rev. Biophysics Biophys. Chem.* **20**: 65-78

Taglietti, V., and Toselli, M. (1988) A study of stretch-activated channels in the membrane of frog oocytes: interaction with Ca^{2+} ions. *J. Physiol.* **407**: 311-328

Trube, G. (1983) Enzymatic dispersion of heart and other tissues, *in:* "Single channel recording," B. Sakmann and E. Neher, eds., Plenum Press, New York, 69-76

Unwin, N. (1989) The structure of ion channels in membranes of excitable cells. *Neuron* **3**: 665-676

Van Duijn, B. (1990) Membrane potential, ion, and their role in the chemotactic response of the cellular slime mold *Dictyostelium discoideum*. Ph.D. thesis, Leiden University, The Netherlands, 87-92

Van Duijn, B., Ypey, D.L., Van der Molen, L.G., and Libbenga, K.R. (1992a) Whole-cell patch-clamp experiments on tobacco protoplasts, *in:* "Progress in Plant Growth Regulation," C.M. Karssen, L.C. van Loon, and D. Vreugdenhil, eds., Kluwer Academic Publishers, Dordrecht, The Netherlands, 668-674

Van Duijn, B., Ypey, D.L., and Libbenga, K.R. (1992) Whole-cell K^+ currents across the plasma membrane of tobacco protoplasts from cell suspension cultures. *Plant Physiol.* (in press)

Wang, Z., Ypey, D.L., and Van Duijn, B. (1992) Inositol trisphosphate-induced hyperpolarization in rat dorsal root ganglion neurons. *FEBS Lett.* **304**: 124-128

Weidema, A.F., Ravesloot, J.H., Panyi, G., Nijweide, P.J., and Ypey, D.L. (1992) A Ca^{2+} dependent K^+ channel in freshly isolated and cultured chick osteoblasts. *Biochim. Biophys. Acta* (in press)

Yamamoto, D., and Suzuki, N. (1987) Blockage of chloride channels by HEPES buffer. *Proc. R. Soc. Lond. B.* **230**: 93-100

Ypey, D.L., and Clapham, D.E. (1984) Development of a delayed outward-rectifying K^+ conductance in cultured mouse peritoneal macrophages. *Proc. Natl. Acad. Sci. (USA)* **81**: 3083-3087

Zimmermann, U., and Steudle, E. (1974) The pressure-dependence of the hydraulic conductivity, the membrane resistance and membrane potential during turgor pressure regulation in *Valonia communis L. J. Membrane Biol.* **16**: 331-352

FLUORESCENT ANALYSIS OF REPLICATION AND INTERMEDIATES OF CHROMATIN FOLDING IN NUCLEI OF MAMMALIAN CELLS

Gaspar Banfalvi

Institute of Biochemistry, Department-I, Semmelweis University Medical School, 1444 Budapest 8, P.O. Box 260, Hungary

INTRODUCTION

The scarcity of evidence regarding chromatin folding is due to the fact that individual chromosomes cannot be seen during their development in the interphase nucleus. Consequently, chromosome structure has been studied primarily in metaphase preparations (Kirsch-Volders et al., 1980; Zang and Back, 1968; Zankl and Zang, 1974), while direct approaches to visualize the topography of chromatin in the nucleus has remained limited (Comings, 1968; Comings, 1980; Cremer et al., 1979; Sperling and Luedtke, 1981; Vogel and Schroeder, 1974) and indirect conclusions were drawn mainly from metaphase studies. This article describes a new approach to the study of chromosome condensation confirming the existence of a flexible folding pattern through a series of transient geometric forms of chromosomes in nuclei of Chinese hamster ovary (CHO) cells. Exponentially growing cells were permeabilized reversibly and nascent DNA was synthesized in the presence of the four dNTPs, dTTP being replaced by biotin-11-dUTP (Banfalvi et al., 1989). Biotinylated nucleotides do not perturb DNA replication (Banfalvi et al., 1989; Blow and Watson, 1987; Hiriyanna et al., 1988; Hunting et al., 1985; Langer et al., 1981; Nakayasu and Bereznay, 1989), but their interference with chromosome folding (Banfalvi et al., 1989, 1990) can be exploited to accumulate intermediates of the condensation process.

MATERIALS AND METHODS

Reagents

2,6-Diamidino-2-phenylindole (DAPI) was the product of Braunschweig Chemie,

Biotechnology Applications of Microinjection, Microscopic Imaging, and Fluorescence, Edited by P.H. Bach *et al.*, Plenum Press, New York, 1993

111

dextran T-150 from Pharmacia, Colcemid from Boehringer, and biotin-11-dUTP, other nucleotides and 1,4-diazobicyclo-(2,2,2)-octane from Sigma. [^{3}H]-Thymidine (77.2 Ci/mmol) was purchased from New England Nuclear and [^{3}H]-dTTP (36.1 Ci/mmol) from Research Production and Utilization of Radioisotopes (Prague).

Isotonic buffer contained 140 mM sucrose, 60 mM KCl, 10 mM Hepes, pH 7.4, 5 mM KPO$_4$, pH 7.4, 5 mM MgCl$_2$, 0.5 mM CaCl$_2$. Antifade medium consisted of 90% w/w 1,4-diazobicyclo-(2,2,2)-octane, 20 mM Tris-Cl, pH 8.0, 0.02% sodium azide and 25 ng/ml DAPI for blue fluorescent staining of DNA. Hypotonic buffer for permeabilization contained 9 mM Hepes, pH 7.8, 5.8 mM dithiothreitol, 4.5% dextran T-150, 1 mM EGTA and 4.5 mM MgCl$_2$.

METHODS

Experimental growth of Chinese hamster ovary cells (CHO-K1 ATCC CCL61) in K-10 medium was tested by flow cytometry (Banfalvi et al., 1989). Permeabilization and reversal of permeabilization, DNA synthesis and its visualization by immunfluorescent microscopy were carried out as described previously (Banfalvi et al., 1984; 1989).

Preparation of Chromosomes and Nuclei

Cells (2 x 10^6) in growth medium for reversal of permeabilization were kept at 37°C for 3 hours and were then treated with 0.1 g/ml Colcemid. After further incubation for 2 hours at 37°C, cells were detached with trypsin and washed with PBS, then kept at 37°C for 10 minutes in 10 ml of 50 mM KCl, 10 mM MgSO$_4$, 3 mM dithiothreitol and 5 mM NaPO$_4$, pH 8.0 swelling buffer. Cells were centrifuged and fixed in 10 ml methanol:glacial acetil acid (3:1).

Visualization of Chromatin Structures

Nuclei and/or chromosome structures were centrifuged at 500xg for 8 minutes, treated twice with 1 ml fixative, and spread over glass slides dropwise from a height of about 30 cm. Slides were air dried, stored at room temperature overnight, rinsed with PBS, and dehydrated using increasing concentrations of ethanol from 70 to 100%.

Dehydrated slides were air dried and mounted in 35 μl antifade medium under 24 x 50 mm coverslips. Blue fluorescence of DAPI staining (25 ng/ml) of DNA was monitored under a Dialux 20 fluorescent microscope (Leitz). Biotinylated DNA was visualized by an immunofluorescent amplification method as described (Banfalvi et al., 1989).

RESULTS AND DISCUSSION

Intermediates of Chromatin Folding in Partially Restored Nuclei

When permeabilized CHO cells were incubated in growth medium containing newborn bovine serum (10%) it was found that the regeneration of cellular membrane

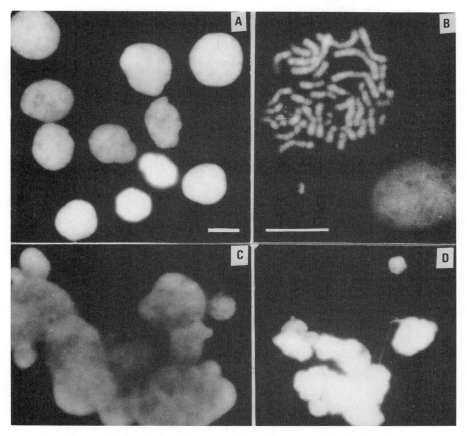

Figure 1. Isolation of interphase nuclei and metaphase chromosomes from exponentially growing CHO cells. A 30 minutes pulse-label of nascent DNA in permeable cells, in the presence of the four dNTPs was followed by the resuspension of cells in growth medium (10% serum) for reversal of permeabilization. Cells were kept at 37°C for 3 hours and then treated with 0.1μg/ml Colcemid. After further incubation (2 hours at 37°C), cells were detached with trypsin and washed with PBS. Swollen cells were centrifuged and fixed in 10 ml methanol:glacial acid (3:1). Preparation of spreads of metaphase chromosomes was the same as that of interphase nuclei. The control included the isolation of nuclei after reversible permeabilization (**A**), without permeabilization (**B**), permeabilization without restoration of membrane integrity (**C** and **D**). Bar 20 μm each.

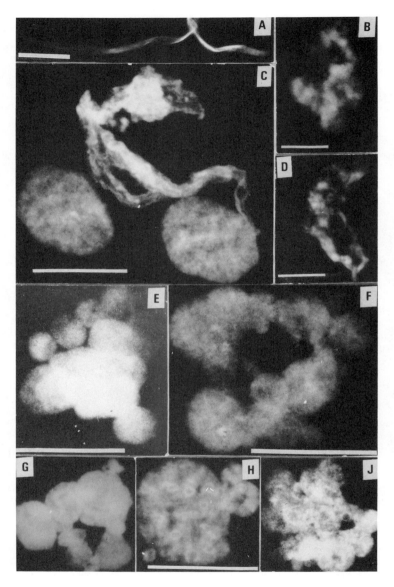

Figure 2. Decondensed forms of chromatin. Pulse label of DNA in permeable cells as in Figure 1 in the presence of dNTPs, dTTP being replaced by 10 μM biotin-11-dUTP followed by the reversal of permeabilization, colcemid treatment and isolation of chromatin structures. Chromatin ribbon (**A**), unfolded, veiled chromatin (**B**), disrupted veil (**C**), supercoiled veil (**D**), chromatosomes (**E, F, G**), semicircled (**H**), looped chromatin (**J**). Bar 20 μm each.

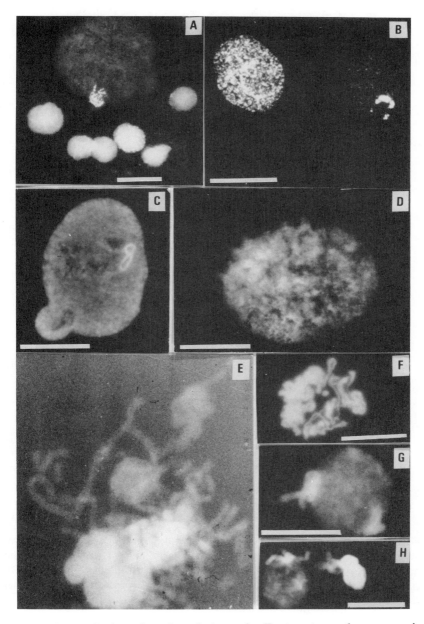

Figure 3. Polarized condensation of chromatin. Treatment was the same as in Figure 2. Immunofluorescent amplification of biotinylated DNA (**A, B**), catenated circles of condensed DNA (**C**), enlarged nucleus with budding out pre-chromosomes (**D**), chromatin bodies turning into fibres (**E**), precondensed chromosomes arranged in a semicircle (**F**), polarized condensation (**G, H**). Bar 20 μm each.

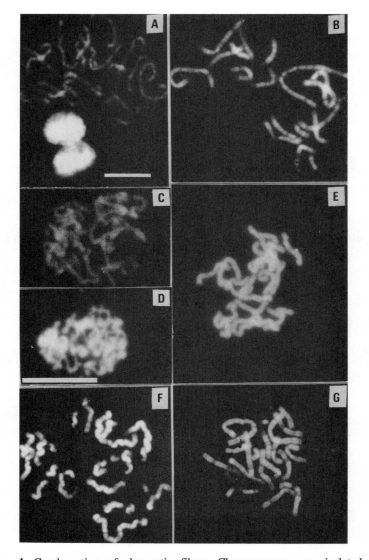

Figure 4. Condensation of chromatin fibres. Chromosomes were isolated from metaphase-arrested CHO cells after labelling with biotin-11-dUTP and partial reversal of permeabilization. Long, decondensed chromosome fibres (**A, B**), connections among condensing chromosomes (**C, D, E**), spiraled forms (**F**), normal chromosomes of metaphase type (**G**). Bar 20 μm each.

gradually slowed down. There was a significant difference in the rate and extent of regeneration between permeable cells kept at high (10%) or low (2%) concentration of newborn bovine serum. The incorporation of [3H]-thymidine showed an opposite tendency to that of [3H]-dTTP incorporation. Simultaneously, the viability of cells was measured during regeneration by trypan blue dye exclusion. After 4 hours of incubation in the presence of a high concentration of bovine serum, approximately 50% of the cells regained their ability to incorporate [3H]-thymidine and only 22% in the presence of low serum concentration. The kinetics of cell regeneration showed that this process slowed down within a few hours. In a control experiment without permeabilization, about 5% of the cells were damaged as a result of experimental manipulation.

Pulse-chase experiments failed to drive biotin label of DNA from interphase to metaphase chromosomes, rather the separation and the expulsion of biotinylated DNA was observed (Banfalvi *et al.*, 1989). Biotinylation as a handicap in chromatin folding was exploited for the accumulation of intermediates of chromosome condensation. This accumulation was enhanced by the incomplete regeneration of the nuclear membrane after permeabilization, allowing the disruption of nuclei and the visualization of intermediates of the chromatin folding process, details of which are not yet known.

Chromatin Structures

Due to the cyclic character of chromatin unfolding (G1-G2 phase) and the condensation, limited to a short period of time (1-2 hours) between prophase and metaphase, exponentially growing CHO cells were treated with colchicine to block the cycle in metaphase (Figure 1). The isolation of chromatin structures from untreated cells results in the two generally known forms: interphase nuclei and metaphase chromosomes (Figure 1b). In another control experiment the isolation of nuclei was carried out without restoration of cellular and nuclear membranes resulting in a sticky mass (Figure 1c and d). Reversal of permeabilization in serum enriched medium was evidenced by thymidine incorporation and trypan blue dye exclusion. The integrity of nuclear membranes after reversal of permeabilization is also demonstrated in Figure 1a with the notable distortion of nuclei reflecting an incomplete regeneration process. Both the incorporation of biotin-11-dUTP inhibiting chromatin folding and the incomplete regeneration of nuclear membranes allowing disruption are necessary criteria to visualize transitional forms of chromosome unfolding and condensation.

In its highly decondensed state, chromatin appears as a long veil-like structure turned around itself, forming the interphase nucleus (Figure 2a-d). An early indication of chromatin decondensation is probably the appearance of round chromatin bodies (Figure 2e-j). The arms and the head portion of the ribboned structure are still distinguishable in Figure 2f. The transition from ribboned to lobular structure is probably caused by the supercoiling of the ribboned chromatin (Figure 2f-j). Chromatin bodies named chromatosomes (Figure 2e) apparently turn to semicircles resembling horseshoe-like arrays (Figure 2h).

Enlargement of Nuclei

The enlargement and incomplete sealing of the nuclear membrane revealed some of the features of the biotinylation of DNA (Figure 3a and b). The polarization of biotin label in enlarged nuclei is an indication that chromatin folding is most effectively prevented in its highly decondensed state (Figure 3a). The separation of biotin label in enlarged nuclei has just been synthesized, blocking the condensation process and leaving the rest of the chromatin unfolded. That this might be the case is demonstrated by the immunofluorescent labelling of newly synthesized DNA forming a premature chromosome, while the rest of the nucleus contains much less fluorescent dye (Figure 3b). It is noteworth that the prematurely formed chromosome, which slipped out of an enlarged nucleus, resembles catenated circles with tetrahedral orientation at the junction region which could form the centromere region of the two chromatids represented by the two circles (Figure 3c). Early forms of chromosomes gradually line up in a semicircle inside the nucleus and bud out in some cases (Figure 3d). These chromosomes obviously consist of two intertwined semicircles, the smaller one pointing outward, the larger semicircle directed inward towards the nucleus. At the beginning of the condensation process chromatin bodies gradually turn to precondensed chromosomes which are lined up (Figure 3e) in a semicircle in the nucleus (Figure 3f). The transition from chromatin bodies to fibres is shown in Figure 3e. Some regions of the chromatin are in decondensed, cotton-like state, others form chromatin fibres and ultimately chromosomes. A defined order of chromosome condensation is indicated by the fact that, while some chromosomes are being formed, others are still in a decondensed state (Figure 3g and h). The observation that condensation takes place at different regions and that the process can be seen at different stages indicates a distinct spatial pattern and a defined order of condensation.

Condensation of Chromosome Fibres

The final stage of chromosome condensation involves the thickening of elongated chromosome fibres as a consequence of further chromatin folding and protein binding. Condensing chromosomes of different length and thickness may represent the final stage of the folding process (Figure 4a-f) leading to the most compact structure known as the metaphase chromosome (Figure 4g). The presence of two lobes in the chromosome network (Figure 4d) supports the notion that chromosomal material can be distinguished as two branches also referred to as chromatin arms in the decondensed, ribboned structure. Condensing chromosomes are interlinked with several other chromosomes through thinner filaments (Figure 4c).

Finally, the simplicity of the chromosome folding immediately suggests a possible mechanism for the reduplication of chromosomes, as follows:-

1. starting with the temporaly defined order of relaxation of individual chromosomes,

2. the separation of decondensed chromatids,

3. reduplication of chromatin bodies,

4. transition from chromatosomes to chromosome fibres and,

5. condensation of chromosome fibres.

REFERENCES

Banfalvi, G., Sooki-Toth, A., Sarkar, N., Csuzi , S., and Antoni, F. (1984) Nascent DNA chains synthesized in reversibly permeable cells of mouse thymocytes. *Eur. J. Biochem.* **139**: 553-559

Banfalvi, G., Wiegant, J., Sarkar, N., and van Duijn, P. (1989) Immunofluorescent visualization of DNA replication sites within nuclei of CHO cells. *Histochemistry* **93**: 81-86

Banfalvi, G., Tanke, H., Raap, A.K., Slats, J., and van der Ploegh, M. (1990) Early replication signals in nuclei of Chinese hamster ovary cells. *Histochemistry* **94**: 435-440

Blow, J.J., and Watson, J.V. (1987) Nuclei act as independent units of replication in *Xenopus* cell-free DNA replication system. *EMBO J.* **6**: 1997-2002

Comings, D.E. (1968) The rationale for an ordered arrangement of chromatin in the interphase nucleus. *Am. J. Hum. Genet.* **20**: 440-460

Comings, D.E. (1980) Arrangement of chromatin in the nucleus. *Hum. Genet.* **53**: 131-143

Cremer, C., Cremer, T., Zorn, C., and Cioreanu, V. (1979) Partial irradiation of Chinese hamster cell nuclei and detection of unscheduled DNA synthesis in interphase and metaphase. A tool to investigate the arrangement of interphase chromosomes in mammalian cells. Hoppe Seyler's Z. *Physiol. Chem.* **360**: 244-245

Hiriyanna, K.T., Varkey, J., Beer, M., and Benbow, M. (1988) Electron microscopic visualization of sites of nascent DNA synthesis by streptavidin-gold binding to biotinylated nucleotides incorporated *in vivo*. *J. Cell Biol.* **107**: 33-44

Hunting, D.J., Dresler, S.L., and Murcia, G. (1985). Incorporation of biotin-labelled deoxyuridine triphosphate into DNA during excision repair and electron microscopic visualization of repair patches. *Biochemistry* **24**: 5729-5734

Kirsch-Volders, M., Hens, L., and Susanne, C. (1980) Telomere and centromere association tendencies in the human male metaphase complement. *Hum. Genet.* **54**: 69-77

Langer, P.R., Waldrop, A.A., and Ward, D.C. (1981) Enzymatic synthesis of biotin-labelled polynucleotides: novel nucleic acid affinity probes. *Proc. Natl. Acad. Sci. USA* **78**: 6633-6637

Nakayasu, H., and Bereznay, R. (1989) Mapping replicational sites in the eukaryotic cell nucleus. *J. Cell Biol.* **108**: 1-11

Sperling, K., and Luedtke, E-K. (1981) Arrangement of prematurely condensed chromosomes in cultured cells and lymphjocytes of the Indian muntjac. *Chromosoma* **83**: 541-553

Vogel, F., and Schroeder, T.M. (1974) The internal order of the interphase nucleus. *Hum. Genet.* **25**: 265-297

Zang, K.D., and Back, E. (1968) Quantitative studies of the arrangement of human chromosomes. I. Individual features in the association pattern of acricentric chromosomes of normal males and females. *Cytogenetics* **7**: 455-470

Zankl, H., and Zang, K.D. (1974) Quantitative studies on the arrangement of human metaphase chromosomes. IV. The association frequency of human acrocentric marker chromosomes. *Hum. Genet.* **23**: 259-265

AUTOFLUORESCENCE IN POTATO TUBER PHELLEM: LOCATION AND TIME COURSE OF APPEARANCE DURING TUBER FORMATION

Theo Hendriks[1], Jan-Wouter van Eck, and Harrie A. Verhoeven

Centre of Plant Breeding and Reproduction Research (CPRO-DLO), Groenendaalsesteeg 1, NL-67 Wageningen, The Netherlands
[1]Current address: Department of Molecular Biology, Dreyenlaan 3, NL-6703 HA Wageningen, The Netherlands

SUMMARY

In cell walls of phellem of potato tubers (*Solanum tuberosum* cv Bintje) a yellow autofluorescence was observed when viewed with blue light (470-490 nm). A similar autofluorescence was also observed in tuber tracheary elements, suggesting that the fluorescence is caused by lignin or lignin-like material. Confocal scanning laser (488 nm) microscopy revealed that in phellem the autofluorescent material is present mainly in the intercellular spaces.

In potato micro-tubers cultured *in vitro* the autofluorescent material was also found in the intercellular spaces of the phellem. Since the development of these micro-tubers is reasonably synchronous, the appearance of the autofluorescence was followed in a time-course study. After tuber formation started, the autofluorescence was observed in intracellular deposits in the epidermis first and in the intercellular spaces between epidermis cells and cuticle shortly thereafter. This suggests that the fluorescent material is secreted from phellem cells via vesicles, and that the secretion is mainly directed towards the distal and radial walls. Later in development, all intercellular spaces between the phellem cells became fluorescent, whereas in full-grown micro-tubers the intracellular deposits were no longer visible. It is concluded that phellem formation in potato tuber cv Bintje is initiated in the epidermis and that the secretion of lignin or lignin-like material is an early step in this process.

Biotechnology Applications of Microinjection, Microscopic Imaging, and Fluorescence, Edited by P.H. Bach *et al.*, Plenum Press, New York, 1993

121

INTRODUCTION

Early during potato tuber development the epidermis of the young tuber is replaced by a periderm. Part of the periderm, the phellem, protects the storage parenchyma from rapid water loss and the intrusion of soil pathogens. The protective function of phellem is mainly ascribed to the presence of suberin in the phellem cell walls. Suberin is a complex heteropolymer composed of phenolic and aliphatic compounds (Kolattukudy, 1980). Periderm formation and suberin deposition also occur upon wounding, and both processes have been extensively studied in wound-healing potato tuber tissue (Barckhausen, 1978; Kolattukudy,1984; Kolattukudy and Espelie, 1985). Much less is known about periderm formation in potato tubers (Cutter, 1978; Peterson *et al.*, 1985).

Phellem cell walls in potato tubers become chemically modified so that they show fluorescence when viewed with ultraviolet or blue light (Peterson *et al.*, 1985). It has been suggested that this is caused by polyphenolics present in the suberized secondary cell wall (Peterson *et al.*, 1978; Espelie *et al.*, 1986).

We investigated the location of the autofluorescence in phellem of potato tuber cv Bintje by confocal scanning laser microscopy to determine whether the autofluorescence is caused by suberin. Phellem of potato micro-tubers cultured *in vitro* was found to greatly resemble phellem in field-grown tubers and the phellem cell walls were also fluorescent when viewed with blue light. For this reason, and their small size, we used micro-tubers to study the time course of the appearance of the autofluorescence by confocal fluorescence microscopy. The formation of these micro-tubers is reasonably synchronous when cultured in the presence of tetcyclacis, an inhibitor of gibberellin biosynthesis (Vreugdenhil, Bindels, Reinhoud, Klocek and Hendriks, in preparation).

MATERIALS AND METHODS

Plant Material

In vitro grown potato micro-tubers were obtained as described in detail elsewhere (Vreugdenhil, Bindels, Reinhoud, Klocek and Hendriks, in preparation). In short, one-node stem pieces from 4 week-old axenic potato plantlets cv Bintje were cultured at $20^{o}C$ in the dark on a solidified (0.8% agar) Murashige and Skoog (1962) medium containing 2% sucrose and 5 μM benzylaminopurine. After 8 days the explants, with stoloniferous shoots developed from their axillary buds, were transferred to a tuber-inducing medium. This medium was similar to the previous except that it contained only one-tenth the normal nitrate concentration, and was supplemented with 8% sucrose, 5 μM benzylaminopurine and 5 μM tetcyclacis. Tuberization of the stoloniferous shoots, visible as a swelling of the region behind the apical hook, started after 4 days in the dark.

Field-grown potato tubers cv Bintje were obtained from a local supermarket.

Microscopy

Confocal fluorescence microscopy was performed on complete micro-tubers or on the cut surface of halved micro-tubers. The confocal scanning laser microscope was a BioRad MRC 500 mounted on a Zeiss ICM 405 inverted microscope equipped with

Neofluar objectives. Objects were put on thin object glasses and a little water was added to improve contact and to prevent dehydration. Excitation was at 488 nm with 1.25 mW laser power. Emitted light was filtered through a BHS barrier filter, transmitting only fluorescence above 515 nm. Thickness of optical sections was less than 1 μm, as determined by measuring fluorescent latex beads of known size.

Fixation of tuber tissue and staining for lipids associated with suberin using Sudan IV according to Nielsen (1973). Free-hand sections were mounted in glycerol and observed with a Nikon Optiphot-2 microscope equipped with epi-optics (B-1A filter block: excitation filter EX470-490, dichromic mirror DM510 and barrier filter BA520).

RESULTS

Localization of the Autofluorescence

Micro-tubers about three weeks old were initially used for confocal fluorescence microscopic analysis of the autofluorescence in potato tubers developed *in vitro*. From the images shown in Figure 1 it can be seen that in these micro-tubers the phellem walls were fluorescent, as well as the cell walls surrounding lenticels (Figure 1A-C).

The occurrence of the yellow fluorescence marked the transition between the stoloniferous shoot and the tuber (Figure 1D-F). The stoloniferous shoot and also the apex of the tuber showed only a scattered pattern of weak fluorescence. Some distinctly localized fluorescence was found present in the basal cell walls of trichomes (Figure 1H) and the thickened cell walls bordering the pore of guard cells (Figure 1I). In the guard cells, plastids were visible due to their red autofluorescence.

The images of the fluorescence obtained from scanning the micro-tuber surface or the cut surface of a halved micro-tuber indicated that the fluorescence is confined to the intercellular spaces, including the spaces between the outermost cells and the cuticle (Figure 1A-C). In all the images obtained after optical sectioning (not shown) we never detected fluorescence in plane with a cell wall. Similar results were obtained from field-grown tubers (not shown). In addition to the intercellular spaces, however, yellow fluorescence was observed in intracellular deposits (Figure 1B). This we never observed in phellem cells of mature field-grown tubers.

Upon staining sections from micro-tubers with Sudan IV the phellem cell walls were stained red (not shown), indicating the presence of suberin. When viewed with epifluorescence microscopy, both Sudan IV staining and the autofluorescence were observed between the phellem cells, but it was difficult to determine whether they were in the same part of the walls, or whether they occurred collateral (not shown).

Yellow fluorescence, very similar to that seen in phellem, was observed in the few tracheary elements present in both field-grown and *in vitro* cultured tubers (not shown). This strongly suggests that the autofluorescence is caused by the presence of lignin or lignin-like material. The tracheary elements did not stain with Sudan IV.

Time Course of the Appearance of the Autofluorescence

Tuber formation *in vitro* started four days after the explants were transferred to the tuber-inducing medium. The epidermis of the swollen region of a stoloniferous shoot at day

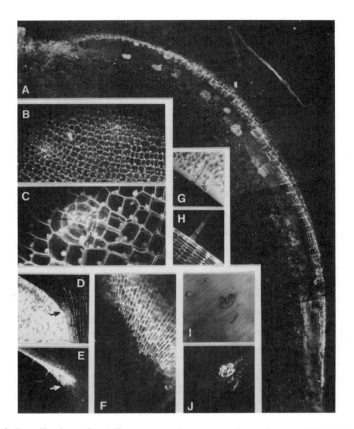

Figure 1. Localization of autofluorescence in potato micro-tubers cv Bintje. Confocal fluorescence images of cut surface of halved micro-tuber (**A**), the surface of the tuber (**B**), detail of **B** showing lenticel (**C**), transition zone (arrow) between tuber and stoloniferous shoot (**E**), detail of **E** (**F**), trichome (**H**), stoma on the stoloniferous shoot (**J**) (plastids in the guard cells are visible due to their red autofluorescence). Images shown in (**D**), (**G**), and (**I**) are transmission microscopic images of **E**, **H**, and **J** respectively. The micro-tubers used were about three weeks old.

4 showed a fluorescence pattern similar to the rest of the stoloniferous shoot (Figure 2a). Two days after the initiation of tuber formation, the yellow fluorescence was observed for the first time in the epidermal cells (Figure 2c). One day later, at day 7, in some regions a rather diffuse fluorescent haze in the distal walls of the epidermal cells was observed, as well as fluorescence in some of the most distal extracellular spaces between two epidermal cells and the cuticle (Figure 2d). Both the intra- and extracellular appearance of the fluorescence were observed first in the medial tuber region then out towards the apex and stolon end.

    Concomitant with the appearance of the fluorescence extracellularly, the surface of the epidermal cells flattened, as could be deduced from the loss of the reflection of the light hitherto observed by transmission microscopy (as can be observed when looking at the cells

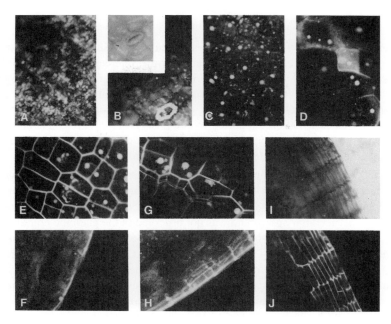

Figure 2. Time course of the appearance of the autofluorescence in potato micro-tubers cv Bintje. (**A-D**) Confocal fluorescence images of the surface micro-tubers at day 4 (**A**), day 5 (**B**), day 6 (**C**), and day 7 (**D**). In (**B**) the insert shows a transmission microscopic image of the stoma. E-H Confocal fluorescence image of (**E, G**) the surface of the tuber and (**F, H**) the cut surface of a halved micro-tuber (**E, F**) at day 10 and (**G, H**) day 14. (**I, J**) Transmission microscopic and fluorescence image of the cut surface of a 4 weeks old micro-tuber.

surrounding the lenticel in the insert of Figure 2b). The intracellular deposits containing the fluorescent material could now also be observed by transmission microscopy (not shown).

About one week after the initiation of tuber development, all the distal extracellular spaces between epidermal cells and cuticle were fluorescent. By that time, in the medial region of the tuber, fluorescence was also observed in some of the radial intercellular spaces between epidermal cells (Figure 2e and f). Thereafter, the fluorescence appeared in the more proximal cells and subsequently in the intercellular spaces, first between these cells and the epidermal cells and followed by the radial intercellular spaces (cf. Figure 2g and h). Finally, after about 4 weeks, full-sized micro-tubers, i.e. 4-5 mm in diameter, had developed 8-10 layers of phellem cells of which the intercellular spaces were fluorescent, whereas in most cells the intracellular deposits were no longer visible (Figure 2i and j). The number of phellem cells in one tier in these micro-tubers was very similar to those observed in field-grown tubers (not shown).

DISCUSSION

The results indicate that the autofluorescence observed in phellem of potato tuber cv Bintje when viewed with blue light is located in the intercellular spaces, and that the autofluorescence can also be found present in intracellular deposits during phellem formation. The appearance of the autofluorescence in intracellular deposits, one day before its appearance in the intercellular spaces, suggests that the fluorescent material is secreted via intracellular vesicles. This is supported by the observation that once phellem formation was completed the intracellular deposits were no longer clearly detectable. The release of the fluorescent material was directed towards the distal and radial walls.

As for the nature of the autofluorescent material, our results show that the autofluorescence is present both in phellem and tracheary elements, suggesting that it is lignin or lignin-like material as in suberin. It is not clear why the autofluorescence is restricted to the intercellular spaces and was not observed in the phenolic domain of suberin present in the walls (as indicated by Sudan IV staining), nor in the lignin present in middle lamella and primary cell wall (Vogt *et al.*, 1983). One possibility may be that the autofluorescence we observed reflects a different degree of polymerization and/or cross-linking of the phenolic compounds. The fact that the autofluorescence is characteristic for the material in the cytoplasmic vesicles may support this. In the middle lamella and primary cell wall the lignin is interspersed between, and cross-linked with, other cell wall material, whereas in suberin the phenolic domain becomes esterified with aliphatic compounds.

Potato tuber periderm has been reported to be initiated by divisions in the epidermis, hypodermis or in both, giving rise to a lateral meristem called phellogen (Cutter, 1978). Phellem cells are considered to be formed from the phellogen towards the outside. The confocal fluorescence images we obtained showed that phellem cells in potato tubers cv Bintje are in tiers with the outermost cells, indicating that phellem formation is initiated in the epidermis only and not in the hypodermis. Furthermore, as shown for the micro-tubers, periclinal cell divisions in phellem formation started before the tuber had reached its final size. Nonetheless, all phellem cells were in tiers with what used to be epidermal cells, suggesting that anticlinal divisions in phellem cells, necessary to keep up with the growth of the tuber, are highly coordinated in that they occur in all cells derived from one epidermal cell by periclinal divisions. These observation are difficult to reconcile with the assumed role of a phellogen in phellem formation.

Autofluorescence is looked upon by many researchers as a nuisance, disturbing the signal obtained by fluorescently labelled antibodies. We think that the results from this study indicated that autofluorescence can be very useful, in particular in combination with confocal scanning laser microscopy, allowing localization of the autofluorescence without sectioning (provided the object is not too big or the autofluorescence is located near the surface). Even though we do not know the exact cause of the autofluorescence, it permitted us to study the formation of potato tuber phellem with the least possible interference with this developmental process.

REFERENCES

Barckhausen, R. (1978) Ultrastructural changes in wounded plant storage tissue cells, *in:* "Biochemistry of wounded plant tissues," G. Kahl, ed., de Gruyter and Co., Berlin, 1-42

Bradford, M.M. (1976) A rapid and sensitive method for quantitation of microgram quantities of protein utilizing the principle of protein-dye binding. *Anal. Biochem.* **72**: 248-254

Cutter, E.G. (1978) Structure and development of the potato plant, *in:* "The potato crop," P.M. Harris, ed., Chapman and Hall, London, 70-152

Espelie, K.E., Franceschi, V.R., Kolattukudy, P.E. (1986) Immunocytochemical localization and time course of appearance of an anionic peroxidase associated with suberization in wound-healing potato tuber tissue. *Plant Physiol.* **81**: 487-492

Kolattukudy, P.E. (1980) Biopolyester membranes of plants: Cutin and suberin. *Science* **208**: 990-1000

Kolattukudy, P.E. (1984) Biochemistry and function of cutin and suberin. *Can. J. Bot.* **62**: 2918-2933

Kolattukudy, P.E., and Espelie, K.E. (1985) Biosynthesis of cutin, suberin, and associated waxes, *in:* "Biosynthesis and biodegradation of wood components," T. Higuchi, ed., Academic Press, New York, 161-207

Murashige, T., and Skoog F. (1962) A revised medium for rapid growth and bioassays with tobacco tissue cultures. *Physiol. Plant.* **15**: 473-479

Nielsen, N.K. (1973) A quick microtechnique for inspection of potato periderm or wound periderm formation. *Potato Res.* **16**:180-182

Peterson, R.L., Barker, W.G., and Howarth, M.J. (1985) Development and structure of tubers, *in:* "Potato physiology," P.H. Li, ed., Academic Press, Orlanda, 123-152

Peterson, C.A., Peterson, R.L., and Robards, A.B. (1978) A correlated histochemical and ultrastructural study of the epidermis and hypodermis of onion roots. *Protoplasma* **96**: 1-21

Vogt, E., Schonherr, J., and Schmidt, H.W. (1983) Water permeability of periderm membranes isolated enzymatically from potato tubers (*Solanum tuberosum* L.). *Planta* **158**: 294-301

FLOW CYTOMETRIC ASSESSMENT OF
GENTAMICIN NEPHROTOXICITY IN
ESTABLISHED RENAL CELL LINES

H. Mc Glynn, E. Healy, T. Bedell, and M. P. Ryan

Department of Pharmacology, University College Dublin
Belfield, Dublin 4, Ireland

INTRODUCTION

The clinical use of gentamicin is limited by its dose-dependent nephrotoxicity. The renal pathogenesis of gentamicin can be attributed to its selective accumulation within the renal proximal tubular cells. Among the clinical signs of gentamicin nephrotoxicity are the renal wasting of calcium and magnesium, low molecular weight proteinuria and enzymuria.

Cultured renal epithelial cells are being developed as an effective tool in the study of nephrotoxins (Gstraunthaler, 1988). In this investigation the established renal cell lines LLC-PK$_1$, (of proximal tubular origin) and MDCK (of distal tubular/collecting duct origin) were used to investigate gentamicin nephrotoxicity using flow cytometric techniques.

Flow cytometry can be used to measure various functional parameters including cell viability and various intracellular free ion concentrations in single living cells. In this study cell viability, intracellular free Ca^{2+} and H^+ concentrations were measured to provide insights into the mechanisms of gentamicin nephrotoxicity.

METHODS

LLC-PK$_1$ and MDCK cells were obtained from Flow Laboratories, Scotland. Cells were maintained in Dulbeccos modified Eagles medium (DMEM) supplemented with 10% foetal calf serum, 2 mM glutamine and incubated at 37°C in 95% air: 5% CO_2 and grown to confluency in 75 cm^2 Costar flasks.

Biotechnology Applications of Microinjection, Microscopic Imaging, and
Fluorescence, Edited by P.H. Bach *et al.,* Plenum Press, New York, 1993

Prior to flow cytometric analysis on a Becton Dickinson FACStar Plus Flow Cytometer, cells were dispersed using trypsin:EDTA mixture (2:1, 0.5% v/v in DMEM : 0.3% v/v in 0.9% NaCl) to form single cell suspensions. Gentamicin at a concentration of 10^{-2} M was added to the single cell suspensions and incubated for 60 minutes at 37°C.

Viability Determination

Fluorescein diacetate (100 ng/ml) and propidium iodide (100 µg/ml) were added to a suspension of single cells (1×10^{6} cells/ml) and allowed to stand at room temperature for 10 minutes. The cells were analysed by recording forward and orthogonal light scatter, red (630 nm) and green (520 nm) fluorescence.

Intracellular pH Determination

Cells were suspended in modified Eagles medium with Earles salts and 0.85 g/l sodium bicarbonate supplemented with 2% foetal calf serum and 2 mM glutamine. Cells were incubated with SemiNaptNoRhodaFluor (SNARF-l, 10 µM) for 30 minutes at 37°C. A standard curve was constructed using various pH buffers incubated with cells in the presence of 2 µg/ml nigericin. Fluorescence was excited with a 514 nm line of the argon-ion laser and collected at 575 nm and 670 nm. The ratio of the 670 nm: 575 nm fluorescence intensity is proportional to increasing pH.

Intracellular Ca^{2+} Determination

Cell suspensions (1×10^{6} cells/ml) were incubated with Indo-1AM (10 µM) for 30 minutes at 37°C. Fluorescence was excited with a 351 nm line of the argon-ion laser and collected at 405 nm and 530 nm. The ratio of the 530 nm: 405 nm fluorescence intensity is proportional to increasing Ca^{2+} concentrations.

RESULTS

Cell viability was assessed by fluorescein diacetate and propidium iodide uptake. This test is based on the principle that fluorescein diacetate is taken up by viable cells where esterases cleave the moiety and it then fluoresces green. Propidium iodide is excluded by viable cells and when taken up by dead or dying cells binds to the nucleic acid and fluoresces red. Viability of both cell types, as judged by these tests, was not significantly affected by a 60 minute treatment with 10^{-2} M gentamicin (Figure 1).

Exposure of cells to gentamicin (10^{-2} M) for 60 minutes (Figure 2) resulted in significant reductions in intracellular pH, as follows; LLC-PK$_1$ from 7.20 ± 0.07 to 6.70 ± 0.04; MDCK from 7.30 ± 0.04 to 7.08 ± 0.04 ($p < 0.005$; n = 5; in both cell types).

Measurement of intracellular free Ca^{2+} (Figure 3) using Indo-1-AM fluorescent ratioing revealed that gentamicin at a concentration of 10^{-2}M for 60 minutes significantly elevated intracellular Ca^{2+} levels in both cell types from control values, given as 100% as follows; LLC-PK$_1$ to 117% $\pm$ 5; MDCK to 109%, $\pm$ 3 ($p < 0.01$; n = 3; in both cell types).

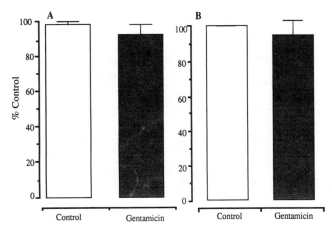

Figure 1. Flow cytometric estimation of cell viability in (**A**) LLC-PK$_1$ and (**B**) MDCK cells treated with gentamicin 10^{-2} M for 60 minutes. Results are expressed as mean ± s.e.m., n=3.

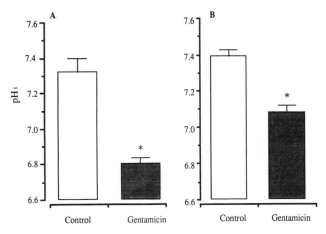

Figure 2. Flow cytometric estimation of intracellular pH in (**A**) LLC-PK$_1$ and (**B**) MDCK cells treated with gentamicin 10^{-2} M for 60 minutes. Results are expressed as mean ± s.e.m., n=3. * $p \leq 0.005$.

DISCUSSION

In this study, it was demonstrated that gentamicin, a known nephrotoxin produced alterations in intracellular pH and Ca^{2+}. Previous results from this laboratory indicated that gentamicin inhibited ^{45}Ca fluxes in brush-border membrane vesicles and primary proximal tubular cells prepared from the rat renal cortex (Godson and Ryan, 1988; McGlynn and Ryan, 1990). The present study indicates that the membrane mediated events associated with gentamicin-induced nephrotoxicity are early manifestations occurring prior to any loss

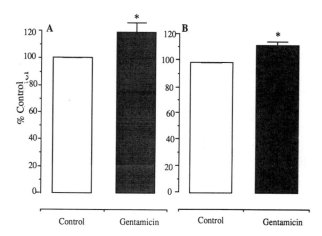

Figure 3. Flow cytometric estimation of intracellular calcium in (**A**) LLC-PK$_1$ and (**B**) MDCK cells treated with gentamicin 10^{-2} M for 60 minutes. Results are expressed as mean ± s.e.m., n=3. * p ≤ 0.01.

of cell viability. Our findings support those of the literature that the primary event in the pathogenesis of gentamicin nephrotoxicity is a rise in intracellular calcium (Holohan *et al.*, 1988), and also with those of Trump *et al.*, (1984) that ion deregulation involving intracellular Na^+, H^+, K^+, Ca^{2+} and Mg^{2+} ions may represent the fundamental lesion in the development of proximal tubule epithelial injury. Flow cytometric analysis is a powerful technique which may be used to elaborate molecular mechanisms of cellular toxicity.

ACKNOWLEDGEMENTS

This work was supported by BioResearch Ireland and by the BRIDGE Programme of E.C. Contract BIOT-CT-910266. The technical assistance of Mr. Roddy Monks is gratefully acknowledged.

REFERENCES

Godson, C., and Ryan, M.P. (1988) Investigation of gentamicin, verapmil and magnesium interactions on calcium uptake in rat renal brush border vesicles. *Br. J. Pharmacol.* **95**: 578

Gstraunthaler, G. (1988) Epithelial cells in tissue culture. *Renal Physiol. Biochem.* **11**: 1-42

Holohan, P.D., Sokol, P.P., Ross, C.R., Coulson, R., Trimble, M.E., Laska, D.A., and Williams, P.D. (1988) Gentamicin-induced increases in cytosolic calcium in pig kidney cells (LLC-PK$_1$). *J. Pharmacol. Exp. Ther.* **247**: 349-354

Mc Glynn, H., and Ryan, M.P. (1990) Gentamicin nephrotoxicity in rat renal proximal tubular cells. *Toxicol. Letts.* **53**: 197-200

Trump, B.F., Jones, R.T., and Heatfield, B.M. (1984) The biology of the kidney. *Adv. Modern Environ. Toxicol.* **7**:273-288

THE USE OF THE FLUORESCENT PROBE FURA-2
FOR INTRACELLULAR FREE CALCIUM
MEASUREMENTS: SOME METHODOLOGICAL
ASPECTS

Anneke Wiltink[1,2], Arnoud van der Laarse[3], Nel P.M. Herrmann-Erlee[2], Joke M. van der Meer[2], and Dirk L. Ypey[1]

[1]Department of Physiology and Physiological Physics
[2]Department of Cell Biology and Histology, and
[3]Department of Cardiology, P.O. Box 9604, NL-2300 RC Leiden, The Netherlands

INTRODUCTION

Fluorescent probes can be used to monitor a variety of free ion concentrations in the cytoplasm of living cells. Amongst the well-known fluorescent probes to estimate the intracellular free calcium concentration ($[Ca^{2+}]_i$) is the tetracarboxylate calcium indicator Fura-2 (Grynkiewicz et al., 1985; Tsien et al., 1985; Tsien, 1988). This probe is widely used in many different cell types to monitor $[Ca^{2+}]_i$ (Malgaroli et al., 1987). Fura-2 has the advantage that its excitation spectrum shifts upon the binding of calcium. This property enables measurement of $[Ca^{2+}]_i$ independent of the intracellular concentration of the dye by using the ratio of fluorescence intensities measured at two wavelengths ("ratio method") (Grynkiewicz et al., 1985). Fluorescence signals of populations of cells can be monitored with the use of spectrofluorometric techniques. Microfluorescence techniques are used to monitor $[Ca^{2+}]_i$ at the single cell level (Tsien and Harootunian, 1990). Although the measurement of Fura-2 fluorescence is easily accessible to a large group of researchers, the interpretation of the observations should be carried out with care. During the last few years many publications have dealt with theoretical as well as practical problems in the use of these techniques (for a review, e.g. see Moore et al., 1990). In addition to calibration problems of the fluorescence signals, the researcher may have to deal with problems concerning loading procedures, leakage of the dye, compartmentalization, binding of the dye to immobile intracellular sites, incomplete hydrolysis of the membrane-permeant

Biotechnology Applications of Microinjection, Microscopic Imaging, and Fluorescence, Edited by P.H. Bach et al., Plenum Press, New York, 1993

Fura-2-acetoxymethyl ester, and photobleaching of the dye during the experiment. An additional problem may be that intracellular Fura-2 contributes to the total cytosolic calcium buffering capacity, and the fluorescence signals may be distorted by the "inner filter effect" (Moore *et al.*, 1990). In the present paper, we discuss some of these problems in relation to our work with bone cells (Ypey *et al.*, 1992; Wiltink *et al.*, 1992), including problems of calibration, dye loading into cells, and dye loss from cells.

METHODS

Foetal Rat Osteoblast Cultures

Osteoblastic cells were isolated from calvaria of 20-day-old rat embryos as described earlier (Herrmann-Erlee *et al.*, 1988). Cell suspensions obtained by the second and third collagenase-treatment were mixed and centrifuged twice at 274 x g for 3 minutes. Cells were diluted in culture medium (CM) consisting of α-MEM (Difco) with 10% foetal calf serum (Gibco), 85 μg/ml gentamicin and 5 mM glucose (pH = 7.2). Aliquots of 100 μl cell suspension were plated on glass cover slips (35 x 10 mm for spectrofluorometric experiments; diameter 24 mm for microfluorescence) in Petri dishes at a final concentration of 1.25 x 10^5 or 2.5 x 10^5 cells/ml. After 2 hours nonadherent cells were removed and CM was added to start culturing until confluency was reached after a few (5) days. After 24 and 72 hours of culture, the medium was replaced by α-MEM containing 5% foetal calf serum, 85 μg/ml gentamicin and 5 mM glucose.

Fluorescence Measurements

Spectrofluorometric experiments were made on confluent monolayers of osteoblastic cultures (Figure 1A) after 2-4 days of culture. The cultures, attached to glass coverslips, were rinsed with a standard extracellular solution (S-ECS) containing: 145 mM NaCl, 5 mM KCl, 1 mM MgCl2, 1 mM CaCl2, 10 mM HEPES, 5 mM glucose (pH 7.2). The cover slip was placed in a quartz glass cuvette and cells were loaded with Fura-2 by incubation for 30 minutes at 37°C in S-ECS containing 2 μM Fura-2/AM (Sigma Chemical Company, St. Louis, U.S.A). After the loading procedures, the cultures were rinsed twice with S-ECS. Experiments were made at either 20°C or 37°C. In part of the experiments performed at 37°C, 2.5 mM probenecid (Sigma) was present during loading and during the fluorescence measurements. Probenecid, which is an organic anion transport blocker, was used to prevent Fura-2 loss from the cells and to reduce compartmentalization (Di Virgilio *et al.*, 1990). Fura-2 loss is defined as the reduction in the fluorescence intensity recorded at 510 nm emission while the excitation wavelength was 360 nm. Since the fluorescence intensity at this excitation wavelength is independent of Fura-2 binding to Ca^{2+}, the intensity recorded at this excitation wavelength can be used as a measure of the total quantity of Fura-2 present in the optical pathway.

Measurements of the intracellular calcium ion concentration ($[Ca^{2+}]_i$) of the monolayers of foetal rat osteoblasts were performed using the ratio method with dual-wavelength excitation (Figure 1B and plate 2). The equipment consisted of a Perkin-Elmer LS-3 fluorescence spectrometer with a Xenon light source. Software (Perkin-Elmer Ltd) run on a PC-XT computer was used to control the excitation monochromator and to acquire the digitized emission values after correction for the dark currents. Excitation wavelengths were 340 and 380 nm, and the emission wavelength was 510 nm. The slit width of the monochromator was 10 nm. The ratio value was obtained by dividing the background-corrected fluorescence intensity (F) at 340 nm excitation (F340) by the background-corrected fluorescence intensity at 380 nm excitation (F380). The background fluorescence, measured before the loading procedures, consisted of autofluorescence of the cells, the media, the glass coverslip, and the cuvette. [Ca^{2+}]$_i$ (in nM) was calculated with the formula described by Grynkiewicz et al. (1985):

$$[Ca^{2+}] = K_d . \beta . \{(R-R_{min})/(R_{max}-R)\} \qquad (1)$$

where K$_d$ is the dissociation constant of Fura-2 binding to Ca^{2+} (224 nM; Grynkiewicz et al., 1985), ß is the maximal fluorescence at 380 nm (F380$_{max}$) divided by the minimal fluorescence at 380 nm (F380$_{min}$), and R is the ratio of two fluorescence intensities, F340/F380. The values R$_{max}$, R$_{min}$, F380$_{max}$ and F380$_{min}$ were obtained by in vivo calibration at the end of each experiment (Moore et al., 1990). The calcium ionophore

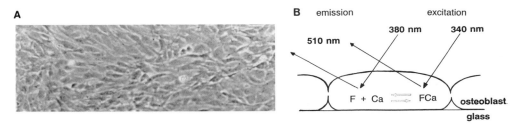

Figure 1. Foetal rat osteoblasts. **A.** Microscopic photograph of a confluent monolayer of foetal rat osteoblasts grown in primary culture for 4 days. The osteoblasts were isolated from calvaria of 20-day-old rat embryos. Cell size is approximately 30 μm. **B.** Schematic drawing of an osteoblastic cell in a monolayer loaded with the fluorescent probe Fura-2 (F). Emission of Fura-2 bound to calcium is maximal if excited by 340 nm ultraviolet light, whereas emission of unbound Fura-2 is maximal if excited by 380 nm ultraviolet light. See also colour plate 2.

ionomycin (Sigma Chemical Company, St. Louis, U.S.A) was used, at a final concentration of 2.5 μM, to saturate the intracellular Fura-2 with calcium in order to obtain R$_{max}$ and F380$_{min}$. Subsequently, the calcium chelator EGTA was used at a final concentration of 20 mM to dissociate intracellular calcium from Fura-2 in order to obtain R$_{min}$ and F380$_{max}$ values. Fluorescence intensities, ratio values and [Ca^{2+}]$_i$ data are expressed as means ± standard deviations for a given number of experiments (n). Statistical analysis was performed using the software program FigP (Biosoft). Differences were considered significant if p-values were less than 0.05.

In some experiments [Ca^{2+}]$_i$ was measured in single cells of nonconfluent monolayers (1-2 day old cultures), attached to the glass cover slips (Ince et al., 1985), using a microfluorescence set-up at 20°C. The equipment consisted of an inverted fluorescence microscope (Carl Zeiss Inc. Axiovert 35) with a 40 x Achrostigmat objective and a Xenon light-source (Carl Zeiss Inc.), a filter wheel with excitation filters (340nm and 380 nm) and filter control unit (Sutter Instrument Company), a dichroic mirror (430 nm), an emission filter (510 nm), and an intensified charge coupled device (CCD)-camera (HCS Vision Technology BV). Software (TIM/IRIAS, Difa) run on a PC-AT compatible computer equipped with an I/O card and a frame-grabber card (PCVISIONplus-AT, Imaging Technology Inc.) was used to control the filter wheel, to acquire the images at 340 nm and 380 nm, and to calculate the ratio images. Each image was built up of 256 x 256 pixels. Each pixel had a value between 0 and 256. After grabbing the images at 340 nm and 380 nm excitation, the dark current and the background fluorescence were subtracted. Templates were applied to the images by setting the pixels outside the cells under a certain threshold to a value of zero (340 nm image) or one (380 nm image). The ratio was obtained by dividing the 340 nm image (F340) by the 380 nm image (F380), followed by multiplication with 99 to fit into the scale (0-256) available. $[Ca^{2+}]_i$ was calculated as was described above.

REVIEW OF SOME POTENTIAL ERROR SOURCES

Three major sources of error in the determination of $[Ca^{2+}]_i$ may be found in the calibration procedure, in the loading procedure, and in the experimental procedure as a result of loss of Fura-2 from the cells, respectively.

Calibration Problems

The calculation of $[Ca^{2+}]_i$ requires the calibration of the fluorescence signals. Therefore, the values of R_{max}, R_{min}, K_d and ß have to be determined (equation [1] above). One method is the *in vitro* calibration procedure. During this procedure, calibration using pentapotassium-salt solutions of Fura-2 is performed in solutions resembling the intracellular solution, and *in vitro* calibration curves are determined by measuring fluorescence at different defined $[Ca^{2+}]_i$. Another method is the *in vivo* calibration at the end of each experiment. During this procedure, calibration is performed intracellularly, and usually only R_{max}, R_{min}, and ß are determined. For calculation of $[Ca^{2+}]_i$, the value of K_d reported by Grynkiewicz et al.(1985) is used. Problems concerning correct calibration of fluorescence measurements using the ratio method have been described recently by many authors (Li *et al.*, 1987; Blatter and Wier, 1990; Lattanzio, 1990; Moore *et al.*, 1990; Roe *et al.*, 1990; Williams and Fay, 1990; Groden *et al.*, 1991; Lattanzio and Bartschat, 1991; Uto *et al.*, 1991).

Several authors have reported that the absorption and emission spectra as well as the K_d of Fura-2 can shift upon changes in polarity, viscosity, ionic strength, pH, and temperature of the solutions (Lattanzio, 1990; Roe *et al.*, 1990; Groden *et al.*, 1991; Moore *et al.*, 1990; Ganz *et al.*, 1990; Poenie, 1990; Uto *et al.*, 1991; Lattanzio and Bartschat, 1991). The use of the formula developed by Grynkiewicz *et al.* [1985; equation (1)] is based on the assumption (Grynkiewicz *et al.*, 1985; Moore *et al.*, 1990) that:-

a) the K_d of Fura-2 in the cell equals the K_d in calibration solutions,

b) the dye is totally de-esterified, and

c) the spectral properties of the dye in the cell equal those in the calibration solution.

It is apparent, therefore, that the uncertainty in $[Ca^{2+}]_i$ may be considerable if $[Ca^{2+}]_i$ is calculated using R_{max}, R_{min}, ß and K_d obtained from *in vitro* calibration. On the other hand, *in vitro* calibration curves can be useful to periodically inspect the equipment and the quality of the dye. During ageing of a mercury light source or after replacement of the light-source, the excitation spectrum may be shifted. The dye may be damaged during lengthy or inappropriate storage. If the *in vitro* calibration curve is not reproducible, this may be a warning to the experimenter.

In vivo calibration at the end of the experiment is usually performed using calcium ionophores like ionomycin or 4-Br-A-23187, and calcium chelators such as EGTA. Finally, $MnCl_2$-solutions can be added to quench the Fura-2 signal in the cytoplasm to assess the amount of compartmentalized Fura-2. Roe *et al.* (1990) described a method to determine compartmentalization using digitonin and Triton-X100. Depending on the cell type, additional ionophores or compounds such as inhibitors and uncouplers of oxidative phosphorylation may have to be used during *in vivo* calibration (Li *et al.*, 1987; Williams and Fay, 1990). Care should be taken that the ionophores are properly solubilized. In the case of ionomycin the strong pH sensitivity of the compound can be a problem (Williams and Fay, 1990). During our experiments, *in vivo* calibration experiments yielded variable calibration values and the dissociation constant (K_d) calculated from the *in vitro* calibration curves was much lower than expected from the value determined by Grynkiewicz *et al.* (1985). Therefore, in our experiments we considered calculated $[Ca^{2+}]_i$ values to be only indications of real $[Ca^{2+}]_i$ values. If calibration procedures yield variable results, it is recommended to report ratio values and to consider relative changes in ratio values. In our view, it is always useful to report calibration values in addition to calculated $[Ca^{2+}]_i$ in a publication, since the reader will then be able to evaluate the results more accurately. It is important to keep in mind that $[Ca^{2+}]_i$ calculated with incorrect calibration values may result in an overestimation of the real effects during the experiment. If *in vivo* calibration yields incorrect R_{max} and R_{min} values, in most cases the real R_{max} and R_{min} values will not be reached during the calibration procedure. This will result in too small R_{max} values and too large R_{min} values. The dynamic range of the dye is then underestimated. Small changes in ratio values will then be considered erroneously as much larger changes in $[Ca^{2+}]_i$.

Here we illustrate from our own experiments a calibration problem related to the effect of bleaching occurring during the measurements. Using our microfluorescence set-up, we investigated bleaching of Fura-2 by deliberately displacing the microscopic field of observation at the end of the experiment partly out of the beam of relatively high intensity (unfiltered) excitation light, so that unexposed cells became visible alongside exposed cells. After the addition of ionomycin for the determination of R_{max}, the difference in R between the unbleached and the bleached region became apparent (Figure 2). Therefore, it is recommended to prevent too high intensities of excitation light during continued fluorescence measurements in order to obtain reliable R values. The possible occurrence of bleaching has to be investigated during control experiments.

Problems of Dye Loading

Several authors have described protocols to load cells with Fura-2 (Roe *et al.*, 1990; Moore *et al.*, 1990). In addition to scrape loading, pH shock, ATP permeabilization and microinjection, the cells can be loaded by incubation in solutions containing the membrane-permeant acetoxymethylester of Fura-2 (Fura-2/AM). The loading procedure is crucial for proper measurement of $[Ca^{2+}]_i$. If cells are not properly loaded, the concentration of Fura-2 inside the cells may be too high or too low. If the cells are loaded using Fura-2/AM, compartmentalization of the dye may occur, or the dye may be incompletely hydrolyzed resulting in various incompletely hydrolyzed products that are highly fluorescent but insensitive to calcium (Oakes *et al.*, 1988; Moore *et al.*, 1990).

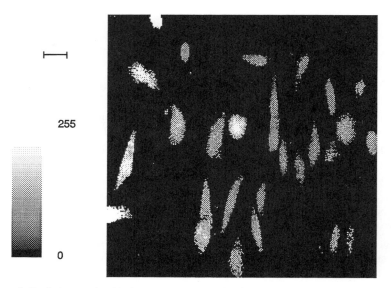

255

0

Figure 2. Ratio image (multiplied by 99 as described in the Methods section) of foetal rat osteoblasts illustrating the effect of bleaching on R_{max} during *in vivo* calibration at the end of the experiment. After bleaching the dye, the culture dish containing foetal rat osteoblasts was deliberately displaced directly before the addition of 2.5 μM ionomycin. The cells at the top and left hand of the image are unexposed cells and reach much higher R_{max} values (mean R = 198) than the cells in the bleached region (right-hand side; mean R = 79). Bar corresponds to 20 μm.

It is advisable to keep the cytosolic Fura-2 concentration below 100 μM, in order to minimize the calcium-buffering capacity of the dye and to prevent distortion of the fluorescence signal via the "inner-filter" effect (Williams and Fay, 1990; Moore *et al.*, 1990). Another reason to keep the cytosolic Fura-2 concentration as low as possible is the fact that hydrolytic products of Fura-2/AM, such as formaldehyde, may exert deleterious effects on cell function. Moore *et al.* (1990) described an efficiency curve for Fura-2/AM loading and a method to estimate the Fura-2 concentration in a single cell. It is important to realize that Fura-2/AM is hydrolyzed inside the cell, and that therefore eventually all Fura-2/AM from the bath may be taken up by the cell. One should, therefore, not only consider the concentration of Fura-2/AM in the bath, but also the volume of the bath and the loading time. In order to assess hydrolysis of Fura-2/AM, spectra of solutions of Fura-2 pentapotassium salts can be compared with spectra of intracellular Fura-2 (Roe *et al.*, 1990; Owen, 1991). Loading the cells at lower temperatures as well as a reduction of the loading time appeared to reduce compartmentalization of the dye (Roe *et al.*, 1990; Moore *et al.*, 1990).

Organic anion transport blockers (probenecid and sulfinpyrazone) have been used to reduce compartmentalization as well as secretion of the dye to the extracellular compartment (DiVirgilio *et al.*, 1990; McDonough and Button, 1989). During our experiments we compared the fluorescence signal of monolayers of osteoblastic cells loaded

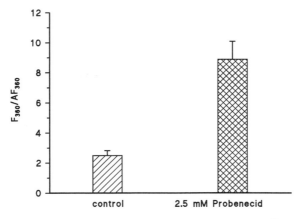

Figure 3. The increase of loading efficiency of Fura-2/AM at 37°C by the organic anion transport inhibitor probenecid (DiVirgilio *et al.*, 1990). The ratio of F360 divided by the autofluorescence of the cells is a measure of the amount of Fura-2 present in the cytoplasm directly after loading. The cellular autofluorescence at 360 nm excitation (AF360) represents the number of cells present in the optical pathway. Data represent mean values ± SEM.

with Fura-2/AM for 30 minutes at 37°C in the presence and absence of probenecid (Figure 3). We compared the fluorescence signal at 360 nm excitation (F360) directly after loading of the cells with the autofluorescence signal of the cells (AF360) at 360 nm excitation. The autofluorescence was recorded in the same cell culture, directly before loading of the cells. If cells were loaded in the presence of probenecid, F360 divided by AF360 was 8.9 ± 5.0 (n = 17), and this value was 2.4 ± 1.2 (n = 13) if cells were loaded without probenecid. Thus, the use of probenecid increased the loading efficiency considerably. This may be explained by the reduction in active secretion of Fura-2 during the loading procedure. Another method to increase the loading efficiency is the use of the dispersing agent Pluronic F-127 (Molecular Probes, U.S.A.) during the loading of the cells with Fura-2/AM.

Loss of Fura-2 from the Cells

Loss of Fura-2 from the cells can be passive as well as active. Passive leakage can occur in damaged or enzymatically isolated cells (Roe *et al.*, 1990). Active secretion has been reported to be inhibited by organic anion transport blockers as described above. Active secretion of the dye and passive leakage present the researcher with several problems. If experiments are performed with cells in solution using a spectrofluorometer, leakage of the dye to the extracellular compartment will result in a gradually increasing background fluorescence. Due to high extracellular [Ca^{2+}], F340 will then increase while F380 decreases, and errors will arise in the recorded ratio values. MnCl$_2$ has been described as useful for quenching extracellular Fura-2 (Roe *et al.*, 1990; McDonough and Button, 1989), but care should be taken since Mn^{2+} ions can block ionic channels specific for Ca^{2+} (Hille, 1992). Another problem induced by leakage of the dye is that the fluorescence signal will decrease as a result of the decrease of the dye concentration in the cytoplasm, causing a

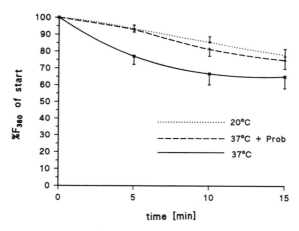

Figure 4. Loss of fluorescence intensity during the experiment recorded as loss in the F360 signal. Data represent mean values ± SEM (n = 6-9). Prob = 2.5 mM probenecid.

decreased signal-to-noise ratio. Performing experiments at temperatures below the physiological range may reduce secretion of the dye to the exterior of the cells (Roe *et al.*, 1990).

During our spectrofluorometric experiments, we recorded the fluorescence signal at an excitation wavelength of 360 nm to estimate the loss of the fluorescence signal due to loss of cells from the coverslip, secretion of Fura-2 to the extracellular compartment, and bleaching of the dye. No bleaching occurred during control experiments as shown by recording ratio values during long (> 15 minutes) periods of time. Ratio values were stable during these control experiments (n = 3), suggesting that no bleaching had occurred. Unfortunately, we could not distinguish between the other two processes, since we did not measure the cells at the single cell level in these experiments.

We compared the loss of F360 during the first 15 minutes after loading in spectrofluorometric experiments performed at 20°C, at 37°C without probenecid in the bath, and at 37°C in the presence of probenecid. A reduced temperature during the experiment as well as the use of probenecid appeared to be equally effective in reducing the loss of F360 signal. Fura-2 loss from foetal rat osteoblasts was 33 ± 16% (n = 6) after 10 minutes if experiments were done at 37°C without probenecid, 19 ± 11% (n = 7) at 37°C in the presence of probenecid and 14 ± 9.7% (n = 8) in the experiments performed at 20°C (Figure 4).

CONCLUSION

The present paper reviewed some potential sources of error in the determination of $[Ca^{2+}]_i$ with the probe Fura-2 with emphasis on calibration, dye loading and dye loss.

It is useful to mention an additional pitfall in the correct determination of $[Ca^{2+}]_i$. Before R values are calculated, background signals have to be subtracted from the F340 and

F380 signals. Background signals arise from cellular fluorescence, from fluorescence due to media surrounding the cells, from drugs added during the experiments, or in general from all possible sources which emit fluorescence light with a wavelength close to the recorded wavelength. Cellular background fluorescence can be determined before the cultures are loaded with Fura-2. In the case of microfluorescence measurements, the actual single cell background can be determined. Then, background measurements, loading procedures and experiments have to be performed without displacing the cells in the field of measurement. Background signals from added drugs should be determined separately. Drugs may induce related problems, since the excitation light or the fluorescence signals of Fura-2 may be distorted by the added compound (Järlebark and Heilbronn, 1992). This has to be examined by the determination of F340 and F380 from Fura-2 in intracellular-like as well as extracellular solutions, after the addition of the particular compounds.

Although the fluorescent probe Fura-2 provides us with the means to monitor $[Ca^{2+}]_i$ fairly easily, the interpretation of the data may be difficult. For example, the experimenter will have to develop proper loading procedures and will have to assess loss of the dye to the extracellular compartment. Furthermore, in calculating $[Ca^{2+}]_i$ from ratio values the various problems described in the present paper will have to be addressed. If calibration results are not fully reliable, results obtained from Fura-2 experiments may be reported as time-dependent R values. Correct calculation of $[Ca^{2+}]_i$ values depends on correct determination of R_{max} and R_{min}. If R_{max} is underestimated and R_{min} is over-estimated, time-dependent $[Ca^{2+}]_i$ changes will be overrated. When reporting R values, part of the problems described here will be circumvented.

ACKNOWLEDGEMENTS

The authors wish to thank Dr. Bert Van Duijn for advice, and Mr. A.J. Schless and Mr. T. Vet of Difa Measuring Systems (Breda, The Netherlands) for cooperation in updating the software for improved image analysis. This work was supported by the Netherlands Organization for Scientific Research (NWO) through grants from the Dutch Foundation for Biophysics.

REFERENCES

Blatter, L.A., and Wier, W.G. (1990) Intracellular diffusion, binding, and compartmentalization of the fluorescent calcium indicators Indo-1 and Fura-2. *Biophys J.* **58**: 1491-1499

Di Virgilio, F., Steinberg, T.H., and Silverstein, S.C. (1990) Inhibition of Fura-2 sequestration and secretion with organic anion transport blockers. *Cell Calcium* **11**: 57-62

Ganz, M.B., Rasmussen, J., Bollag, W.B., and Rasmussen, H. (1990) Effect of buffer systems and pH$_i$ on the measurement of $[Ca^{2+}]_i$ with Fura-2. *FASEB J* **4**: 1638-1644

Grynkiewicz, G., Poenie, M., and Tsien, R.Y. (1985) A new generation of calcium indicators with greatly improved fluorescence properties. *J Biol Chem* **260**: 3440-3450

Groden, D.L., Guan, Z., and Stokes, B.T. (1991) Determination of Fura-2 dissociation constants following adjustment of the apparent Ca-EGTA association constant for temperature and ionic strength. *Cell Calcium* **12**: 279-287

Herrmann-Erlee, M.P.M., van der Meer, J.M., Löwik, C.W.G.M., van Leeuwen, J.P.T.M., and Boonekamp, P.M. (1988) Different roles for calcium and cyclic AMP in the action of PTH: Studies in bone xplants and isolated bone cells. *Bone* **9**: 93-100

Hille, B. (1992) "Ionic channels of excitable membranes", 2nd Edn., Sinauer Associates Inc., Sunderland, Massachusetts, 108-111

Ince, C., van Dissel, J.T., and Diesselhoff-den Dulk, M.M.C. (1985) A teflon culture dish for high magnification microscopy and measurements in single cells. *Pflügers Arch* **403**: 240-244

Järlebark, L., and Heilbronn, E. (1992) Tetrahydroaminoacridine and related compounds interfere with Fura-2 and Indo-1. *Eur J Pharmacol* **225**: 75-77

Lattanzio, F.A. (1990) The effects of pH and temperature on fluorescent calcium indicators as determined with chelex-100 and EDTA buffer systems. *Biochem. Biophys. Res. Commun.* **171**: 102-108

Lattanzio, F.A., and Bartschat, D.K. (1991) The effect of pH on rate constants, ion selectivity and thermodynamic properties of fluorescent calcium and magnesium indicators. *Biochem. Biophys. Res. Commun.* **177**: 184- 191

Li, Q., Altschuld, R.A., and Stokes, B.T. (1987) Quantitation of intracellular free calcium in single adult cardiomyocytes by Fura-2 fluorescence microscopy: calibration of Fura-2 ratios. *Biochem Biophys Res Commun* **147**: 120-126

Malgaroli, A., Milani, D., Meldolesi, J., and Pozzan, T. (1987) Fura-2 measurement of cytosolic free Ca^{2+} in monolayers and suspensions of various types of animal cells. *J Cell Biol* **105**: 2145-2155

McDonough, P.M., and Button, D.C. (1989) Measurement of cytoplasmic calcium concentration in cell suspensions: correction for extracellular Fura-2 through the use of Mn^{2+} and probenecid. *Cell Calcium* **10**: 171-180

Moore, E.D.W., Becker, P.L., Fogarty, K.E., Williams, D.A., and Fay, F.S. (1990) Ca^{2+} imaging in single living cells: Theoretical and practical issues. *Cell Calcium* **11**: 157-179

Oakes, S.G., Martin, H.W.J., Lisek, C.A., and Powis, G. (1988) Incomplete hydrolysis of the calcium indicator precursor Fura-2 Pentaacetoxymethyl Ester (Fura-2 AM) by cells. *Anal. Biochem.* **169**: 159-166

Owen, C.S. (1991) Spectra of intracellular Fura-2. *Cell Calcium* **12**: 385-393

Poenie, M. (1990) Alteration of intracellular Fura-2 fluorescence by viscosity: A simple correction. *Cell Calcium* **11**: 85-91

Roe, M.W., Lemasters, J.J., and Herman, B. (1990) Assessment of Fura-2 for measurements of cytosolic free calcium. *Cell Calcium* **11**: 63-73

Tsien, R.Y., Rink, T.J., and Poenie, M. (1985) Measurement of cytosolic free Ca^{2+} in individual small cells using fluorescence microscopy with dual excitation wavelengths. *Cell Calcium* **6**: 145-157

Tsien, R.Y. (1988) Fluorescence measurement and photochemical manipulation of cytosolic free calcium. *TINS* **10**: 419-242

Tsien, R.Y., and Harootunian, A.T. (1990) Practical design criteria for a dynamic ratio imaging system. *Cell Calcium* **11**: 93-109

Uto, A., Arai, H., and Ogawa, Y. (1991) Reassessment of Fura-2 and the ratio method for determination of intracellular Ca^{2+} concentrations. *Cell Calcium* **12**: 29-37

Williams, D.A., and Fay, F.S. (1990) Intracellular calibration of the fluorescent calcium indicator Fura-2. *Cell Calcium* **11**: 75-83

Wiltink, A., Nijweide, P.J., Herrmann-Erlee, M.P.M., van der Plas, A., van der Meer, J.M., and Ypey, D.L. (1992) Influence of 4-aminopyridine and valinomycin on the intracellular calcium concentration of cultured embryonic rat and chick osteoblasts. *Pflügers Arch.* **420**: R80 (abstract)

Ypey, D.L., Weidema, A.F., Höld, K.M., van der Laarse, A., Ravesloot, J.H., van der Plas, A., and Nijweide, P.J. (1992) Voltage, calcium and stretch activated ionic channels and intracellular calcium in bone cells. *J. Bone Min. Res.* (in press)

STUDIES ON THE EFFECTS OF PEROXISOME PROLIFERATORS AS NOVEL Ca^{2+} MOBILIZING AGENTS USING INDO-1-LOADED HEPATOCYTES

Anton M. Bennett[1], Oliver P. Flint[2], and Gary M. Williams[1]

[1]American Health Foundation, Department of Pathology and Toxicology and New York Medical College, Department of Experimental Pathology, NY 10595, U.S.A.
[2]Bristol-Myers Squibb, Pharmaceutical Research Institute, Syracuse, NY 13221, U.S.A.

INTRODUCTION

A class of chemically diverse rodent hepatocarcinogens has been identified which induces the proliferation of both hepatocytes and peroxisomes, the hepatocellular organelles involved in fatty acid metabolism (Reddy and Rao, 1986; Rao and Reddy, 1987). These agents include pharmaceutical compounds, such as hypolipidemic drugs, as well as industrial plasticizing agents and pesticides. Collectively, these chemicals have been termed peroxisome proliferating agents (PPA). Despite the carcinogenicity of these agents, they fail to interact with DNA (Goel et al., 1985; Schiestl and Reddy, 1990). Therefore, the PPA constitute a distinct class of non-genotoxic hepatocarcinogens. To date, no definitive evidence exists to explain the carcinogenicity or tumour promoting capacity of the PPA. However, the ability of the PPA to induce hepatocyte proliferation, oxidative stress (as a result of peroxisomal proliferation), and to also bind a novel steroid-like receptor (Issemann and Green, 1990), have all been suggested to play an important role in the carcinogenicity of the PPA. It is likely, however, that all of these components play an important role in the carcinogenicity and tumour promoting capacity of the PPA.

In light of these observations, we have addressed the mechanisms of PPA-induced hepatocyte proliferation. The liver typically responds to a PPA by initiating a rapid, but transient, induction of hepatocyte DNA synthesis occurring within the first three days after treatment both *in vivo* (Yeldandi et al., 1989) and in culture (Bieri et al., 1984). This is accompanied by both a hyperplastic and hypertrophic response leading to hepatomegaly

Biotechnology Applications of Microinjection, Microscopic Imaging, and Fluorescence, Edited by P.H. Bach *et al.*, Plenum Press, New York, 1993

143

within days following treatment. In this study, we have investigated the possible operating mechanism(s) which may be involved in the induction of hepatocyte DNA synthesis after treatment of cultured rat hepatocytes with a PPA. We have hypothesized that the PPA may induce hepatocyte DNA synthesis and subsequently hepatocyte proliferation in a manner that may be mechanistically similar to other endogenous growth factors. This hypothesis, referred to as the "growth factor paradigm of peroxisome proliferator-induced hepatocyte DNA synthesis" led us to examine the role of the Ca^{2+} signal transduction pathway in response to the PPA. The role of Ca^{2+} as a pivotal cation in growth control and regulation is well established, and has been reviewed extensively elsewhere (Whitfield *et al.*, 1987). Rapid and transient changes in the free intracellular Ca^{2+} concentration ($[Ca^{2+}]_i$) is a hallmark response in a variety of different cell types responding to growth factors such as Epidermal Growth Factor (EGF), Platelet-Derived Growth Factor (PDGF) and Transforming Growth Factor-alpha (TGF-α) (Moolenaar *et al.*, 1984; Wong *et al.*, 1989). This response leads to the activation of numerous Ca^{2+}-dependent processes which play a critical part in the cascade of events leading to replicative DNA synthesis and cell proliferation. Using the isolated rat hepatocyte model, we have employed the use of the fluorescent Ca^{2+} probe, Indo-1-AM (Grynkiewicz *et al.*, 1985), to investigate whether the hypolipidemic PPA, ciprofibrate and clofibrate, increase hepatocyte $[Ca^{2+}]_i$. These studies have revealed that both ciprofibrate and clofibrate elevate the hepatocyte $[Ca^{2+}]_i$ and that the kinetics of the $[Ca^{2+}]_i$ increase is distinct between ciprofibrate and clofibrate, despite their structural similarities. Furthermore, we have identified that the mechanism of ciprofibrate-induced increases in hepatocyte $[Ca^{2+}]_i$ stems from the inhibition of the endoplasmic reticulum (ER) Ca^{2+}-ATPase, one of the major regulators of intracellular Ca^{2+} stores (Carafoli, 1987).

MATERIALS AND METHODS

Materials

Collagenase (type I) and clofibrate (2-(p-chlorophenoxy)-2-methylpropionic ethyl ester) were obtained from Sigma Chemical Co. (St. Louis, MO). Ciprofibrate ((2-[4-(2,2-dichlorocyclopropyl)-phenoxy]-2-methylpropionic ethyl ester) was a kind gift from Sterling Research Institute, NY. [Arg[8]] vasopressin (VP) was obtained from Boehringer Mannheim (Indianapolis, IN). The fluorescent Ca^{2+} probe, Indo-1-acetomethyl ester (Indo-1-AM) was obtained from Molecular Probes Inc. (Eugene, OR). Remaining chemicals and reagents were obtained from commercial suppliers and were of the highest purity available.

Hepatocyte Isolation and Culture

Hepatocytes were isolated by a two-stage collagenase digestion through intact livers obtained from male F344 rats (200-250 g) as described previously (Williams *et al.*, 1982) with minor modifications. Briefly, rats were anesthetized with phenobarbital sodium (50 mg/kg) and livers perfused via cannulation of the hepatic portal vein, first with a Ca^{2+}-free solution (50 mM HEPES-KOH (pH 7.4), 100 mM NaCl, 5.5 mM D-glucose, 5.4 mM KCl, 4.4 mM KH_2PO_4, 3.3 mM Na_2HPO_4, 15 mM $NaHCO_3$, 0.5 mM EGTA and 0.25 mg/ml insulin) followed by a collagenase digestion solution (50 mM HEPES-KOH (pH 7.6), 100 mM NaCl, 5.5 mM D-glucose, 5.4 mM KCl, 15 mM $NaHCO_3$, 6 mM $CaCl_2$, 0.25 mg/ml insulin and 100 units/ml collagenase type I). Following collagenase digestion, hepatocytes were dissociated and viability assessed by trypan blue exclusion. Freshly isolated rat hepatocyte cultures were then preincubated at 2 x 10^6 cells/ml in Buffer A (modified Krebs-Henseleit buffer containing 10 mM HEPES-KOH, 121 mM NaCl, 4.7 mM KCl, 1.2 mM NaH_2PO_4, 1.2 mM $MgSO_4$, 2.0 mM $CaCl_2$, 5.0 mM $NaHCO_3$, 10 mM ι-glucose and 0.2% (w/v) bovine serum albumin, pH 7.4) at 37°C under an atmosphere of 95% air and 5% CO_2.

Measurement of Hepatocyte $[Ca^{2+}]_i$

Hepatocyte $[Ca^{2+}]_i$ was measured with the single excitation, dual emission fluorescent Ca^{2+} probe Indo-1-AM using an Hitachi F-2000 spectrofluorimeter with monochromator settings at 355 nm (excitation) and 405 nm and 485 nm (emission) detection. Although Fura-2 has become the preferred probe for $[Ca^{2+}]_i$ determinations, Indo-1 is equally suited. In fact, Indo-1 is the probe of choice for measurement of $[Ca^{2+}]_i$ using flow cytometry where only one excitation wavelength can be achieved, ratioed measurements are subsequently obtained by capture of the dual emission spectra using two photomultipier detectors. Furthermore, studies comparing the use of Indo-1 and Fura-2 have found that Indo-1 was less prone to subcellular organelle compartmentalization than Fura-2 (Wahl et al., 1990). More recently, Ca^{2+} probes such as Fluo-3, have been developed (Kao et al., 1989), which allow $[Ca^{2+}]_i$ determinations to be made within the visible wavelength. The disadvantage of Fluo-3 however, is that it does not employ ratio imaging and, consequently, many of the inherent errors incorporated during fluorescent detection cannot be eliminated. Nonetheless, the use of Fluo-3 does represent an extremely applicable means of measuring $[Ca^{2+}]_i$. It appears, therefore, that these probes are equally suited for measuring $[Ca^{2+}]_i$, the decision as to which probe to use is solely dependent upon each specific experimental situation.

After pre-incubation of hepatocyte suspension cultures in Buffer A, Indo-1-AM (5 μM) was added to 2 x 10^6 cells/ml for approximately 30 minutes at 37°C. Cells were then washed twice by spinning down at 50 x g for 2 minutes and resuspended in fresh Buffer A. To allow for complete de-esterification of Indo-1-AM, cells were subsequently incubated for at least a further 15 minutes at 25°C (Scanlon et al., 1987). For the measurement of hepatocyte $[Ca^{2+}]_i$ in the presence of extracellular Ca^{2+} (2 mM $CaCl_2$), hepatocyte suspensions (2 ml) were diluted to 1.0 x 10^6 cells/ml in Buffer A and preincubated for approximately 5 minutes at 37°C prior to the addition of either VP, ciprofibrate or clofibrate. For experiments carried out in the absence of extracellular Ca^{2+}, hepatocytes (1 x 10^6 cells/ml) were preincubated in Buffer A for 5 minutes at 37°C followed by the addition of 1 mM EGTA. After a stable fluorescent baseline had been achieved, cells were then challenged with the PPA. Addition of 1 mM EGTA did not alter the fluorescence signal, indicating minimal dye leakage was occurring. After each hepatocyte suspension culture had been challenged with either VP or a PPA, $[Ca^{2+}]_i$ calibration values were obtained by permeabilization of the hepatocyte suspension with 0.1% Triton X-100 followed by the addition of 10 μl/ml EGTA (0.25 M in 2.0 M Tris-HCl, pH 8-8.5) for the determination of maximum (F_{max}) and minimum (F_{min}) fluorescence, respectively. Calculation of $[Ca^{2+}]_i$ was determined by application of equation (1) (Grynkiewicz et al., 1985). It is important to

$$[Ca^{2+}]_i = K_d(F - F_{min}) / (F_{max} - F) \qquad (1)$$

note that application of the Indo-1 K_d (250 nM) to this equation is only appropriate for experiments carried out at 37°C. Use of K_d values at temperatures other than that from which they are derived has been demonstrated to significantly influence the quantitative determination of $[Ca^{2+}]_i$ (Shuttleworth and Thompson, 1991).

RESULTS

Ciprofibrate and Clofibrate Increase Hepatocyte $[Ca^{2+}]_i$

When isolated rat hepatocytes were challenged with 50 nM VP in the presence of extracellular Ca^{2+}, a rapid and transient increase in hepatocyte $[Ca^{2+}]_i$ was observed (Figure 1a). This response is typical of a receptor-mediated event operating via the generation of inositol-1,4,5-trisphosphate (IP3) (Berridge and Irvine, 1989). When freshly loaded Indo-1 rat hepatocytes are challenged with the PPA, ciprofibrate, at 100 and 200 µM, a rapid increase in hepatocyte $[Ca^{2+}]_i$ is also observed to occur in a concentration-dependent manner, as illustrated in Figures 1b and 1c. The increase in hepatocyte $[Ca^{2+}]_i$ in response to 100 and 200 µM ciprofibrate was approximately 60 and 120 nM respectively. Importantly, however, ciprofibrate was observed to produce a sustained elevation of hepatocyte $[Ca^{2+}]_i$ in the presence of extracellular Ca^{2+} in a manner that is distinct from the transient increase in $[Ca^{2+}]_i$ induced by treatment with VP. In the absence of extracellular Ca^{2+}, ciprofibrate (200 µM) displayed a transient elevation in hepatocyte $[Ca^{2+}]_i$ (Figure 2b), indicating that ciprofibrate was mobilizing intracellular Ca^{2+} stores. When Indo-1-loaded rat hepatocytes were treated with clofibrate, a transient elevation of hepatocyte $[Ca^{2+}]_i$ was observed in the presence of extracellular Ca^{2+}, in a concentration-dependent manner at 100 and 200 µM (Figure 3). Clofibrate-induced $[Ca^{2+}]_i$ changes were identical to that of the transient $[Ca^{2+}]_i$ increase produced with VP.

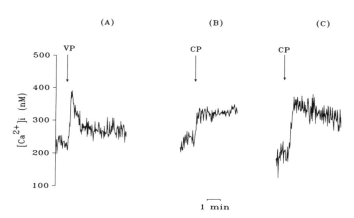

Figure 1. Isolated rat hepatocytes were loaded with Indo-1-AM (5 µM) and $[Ca^{2+}]_i$ was determined as described in Materials and Methods in response to (**A**) 50 nM [arg[8]] vasopressin (VP), (**B**) 100 µM, and (**C**) 200 µM ciprofibrate (CP) in the presence of extracellular Ca^{2+}.

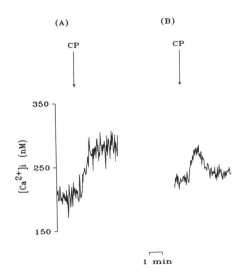

Figure 2. Treatment of isolated rat hepatocytes with 200 μM CP in (**A**) the presence of extracellular Ca^{2+} (2 mM CaCl2), and (**B**) absence of extracellular Ca^{2+} (1 mM EGTA).

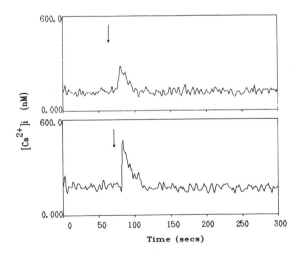

Figure 3. Response of Indo-1-loaded rat hepatocytes following treatment with 100 μM (top panel) and 200 μM (bottom panel) clofibrate in the presence of extracellular Ca^{2+} (2 mM CaCl2).

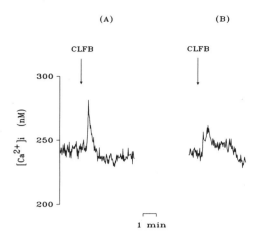

Figure 4. Clofibrate (CLFB) at 200 μM was administered to Indo-1-loaded hepatocytes in (**A**) the presence of extracellular Ca^{2+} (2 mM $CaCl_2$), and (**B**) absence of extracellular Ca^{2+} (1 mM EGTA).

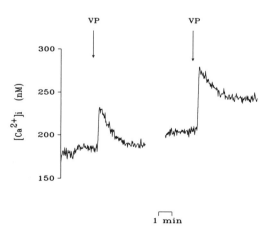

Figure 5. Cultured rat hepatocytes were pre-incubated in Buffer A in the presence of ciprofibrate (200 μM) or DMSO (as control) for 1 hour, after which time hepatocytes were washed and established in fresh Buffer A. VP (50 nM) was then added to (**A**) control cells, and (**B**) ciprofibrate-pretreated cells, in the presence of extracellular Ca^{2+} (2 mM $CaCl_2$).

Clofibrate-induced increases in $[Ca^{2+}]_i$ were also dependent upon extracellular Ca^{2+}, as the observed Ca^{2+} response was suppressed in the absence of extracellular Ca^{2+} (Figure 4b).

The sustained elevation of hepatocyte $[Ca^{2+}]_i$ following treatment with ciprofibrate suggested that alterations in Ca^{2+} sequestration may be operating, as opposed to a receptor-mediated event which classically produces a transient rise in $[Ca^{2+}]_i$. This inference was further supported by evidence indicating that another hypolipidemic PPA, nafenopin, also increased hepatocyte $[Ca^{2+}]_i$ but without generation of inositol phosphates (Ochsner et al., 1990). We have subsequently demonstrated that ciprofibrate does indeed alter intracellular Ca^{2+} sequestration by reducing the ER Ca^{2+}-ATPase activity (Bennett and Williams, 1992). The ER Ca^{2+}-ATPase functions to sequester cytosolic Ca^{2+} into the ER thereby attenuating rises in $[Ca^{2+}]_i$ and simultaneously replenishing the IP_3 releasable Ca^{2+} pool (Kass et al., 1989). Thus, impaired cytosolic Ca^{2+} sequestration will lead to elevated levels of $[Ca^{2+}]_i$. Collectively, these data strongly suggest that ciprofibrate-induced $[Ca^{2+}]_i$ increases occur as a function of ER Ca^{2+}-ATPase inhibition. The nature of the clofibrate-induced Ca^{2+} response, however, remains to be elucidated.

Effect of Ciprofibrate on VP-induced $[Ca^{2+}]_i$ Response

Freshly isolated rat hepatocytes (2×10^6 cells/ml) were incubated in Buffer A at $37^{\circ}C$ for 1 hour with either DMSO, as control, or ciprofibrate ($200 \mu M$). Cells were then washed twice by spinning down at $50 \times g$ for 2 minutes, resuspended in fresh Buffer A and loaded with $5 \mu M$ of Indo-1-AM for 30 minutes at $37^{\circ}C$. Following loading with Indo-1-AM, hepatocytes were diluted to 1×10^6 cells/ml in Buffer A and the fluorescence in response to 50 nM VP recorded in control and ciprofibrate pre-treated hepatocytes. Figure 5a shows the $[Ca^{2+}]_i$ response to 50 nM VP in control hepatocyte suspension cultures, which produce the typical transient rise in $[Ca^{2+}]_i$. Treatment of hepatocytes with ciprofibrate ($200 \mu M$) for 1 hour followed by 50 nM VP challenge did not produce a transient elevation in hepatocyte $[Ca^{2+}]_i$. Rather, a sustained elevation in hepatocyte $[Ca^{2+}]_i$ was observed, indicating an inability to restore basal $[Ca^{2+}]_i$, presumably as a consequence of reduced ER Ca^{2+}-ATPase activity. This supports previous data describing the inhibition of the ER Ca^{2+}-ATPase by ciprofibrate.

CONCLUSIONS

The PPA behave as potent hepatic mitogens and this effect has been suggested to play an important role in the hepatocarcinogenicity and tumour promoting capacity of these agents. We have employed the fluorescent Ca^{2+} probe Indo-1 to the isolated rat hepatocyte model to demonstrate that the hypolipidemic PPAs, ciprofibrate and clofibrate, increase hepatocyte $[Ca^{2+}]_i$. Use of Ca^{2+} probes to investigate changes in $[Ca^{2+}]_i$ has become a well established methodology, which has been applied to a variety of different cell systems. Although this technique has become relatively straightforward, caution should be exercised when establishing calibration data from the ratioed emission outputs to the quantitative determination of $[Ca^{2+}]_i$. Use of the appropriate K_d for the probe is of course imperative and should ideally be applied (cf. equation 1) under the conditions from which they were derived. In particular, temperature and pH appear to be important factors.

The mobilization of hepatocyte $[Ca^{2+}]_i$ was distinct between ciprofibrate and clofibrate, despite their structural similarities. Whereas ciprofibrate produced a sustained increase in hepatocyte $[Ca^{2+}]_i$ in the presence of extracellular Ca^{2+}, clofibrate produced a rapid and transient elevation of $[Ca^{2+}]_i$. The nature of these distinct responses between ciprofibrate and clofibrate remain to be elucidated. Nonetheless, we have established that the ciprofibrate-induced $[Ca^{2+}]_i$ response occurs as a result of the inhibition of the ER Ca^{2+}-ATPase, an essential high affinity Ca^{2+} sequestering enzyme of intracellular Ca^{2+} pools. This observation is further supported by evidence showing that another hypolipidemic PPA, nafenopin, also increases hepatocyte $[Ca^{2+}]_i$, but does so independently of inositol phosphate metabolism. Ciprofibrate was observed to potentiate VP-induced $[Ca^{2+}]_i$ increases, presumably as a result of the inability to sequester and thereby restore basal cytoplasmic Ca^{2+}.

Application of the fluorescent Ca^{2+} probe Indo-1 has allowed us to identify a potentially important mechanism underlying PPA-induced hepatocyte proliferation. However, not all probes necessarily have to fluoresce to be useful; indeed equal credit should be given to the acetomethyl ester derivative which allows the otherwise membrane impermeant probe to be incorporated into the cell. For example, we have applied the use of BAPTA-AM (bis-(o-aminophenoxy)-ethane-N,N,N′,N′–tetracetic acid, an intracellular Ca^{2+} chelator, to investigate the role of ciprofibrate-induced Ca^{2+} increases in the induction of hepatocyte replicative DNA synthesis. These studies have revealed that co-incubation of ciprofibrate with BAPTA-AM reduces the magnitude of hepatocyte DNA synthesis induced by ciprofibrate. Intracellular probes, therefore, can be potentially applied to a variety of biological systems which can subsequently be used as tools to address a spectrum of diverse biological questions.

REFERENCES

Bennett, A.M., and Williams, G.M. (1992) Reduction of rat liver endoplasmic reticulum Ca^{2+}ATPase activity and mobilization of hepatic intracellular calcium by ciprofibrate, a peroxisome proliferator. *Biochem. Pharmacol.* **43**: 595-605

Berridge, M.J., and Irvine R.F. (1989) Inositol phosphates and cell signalling. *Nature* **341**: 197-205

Bieri, F., Bentley, P., Waechter, F., and Staubli, W. (1984) Use of primary cultures of adult rat hepatocytes to investigate mechanisms of action of nafenopin, a hepatocarcinogenic peroxisome proliferator. *Carcinogenesis* **5**(8): 1033-1039

Carafoli, E. (1987) Intracellular calcium homeostasis. *Annu. Rev. Biochem.* **56**: 395-433

Goel, S.K., Lalwani, N.D., Fahl, W.E., and Reddy, J.K. (1985) Lack of covalent binding of peroxisome proliferators nafenopin and Wy-14,643 to DNA *in vivo* and *in vitro*. *Toxicol. Lett.* **24**: 37-43

Grynkiewicz, G., Poenie, M., and Tsien, R.Y. (1985) A new generation of Ca^{2+} indicators with greatly improved fluorescence properties. *J. Biol. Chem.* **260**: 3440-3450

Issemann, I., and Green, S. (1990) Activation of a member of the steroid hormone receptor superfamily by peroxisome proliferators. *Nature* **347**: 645-650

Kao, J.P.Y., Harootunian, A.T., and Tsien, R.Y. (1989) Photochemically generated cytosolic pulses and their detection by Fluo-3. *J. Biol. Chem.* **264**: 8179-8184

Kass, G.E.N., Duddy, S.K., Moore, G.A., and Orrenius, S. (1989) 2,5-Di-(tert)-1,4-benzohydroquinone rapidly elevates cytosolic Ca^{2+} concentration by mobilizing the inositol 1,4,5-trisphosphate-sensitive Ca^{2+} pool. *J. Biol. Chem.* **264**: 15192-15198

Moolenaar, W.H., Tertoolen, L.G.J., and de Laat, S.W. (1984) Growth factors immediately raise cytoplasmic free Ca^{2+} in human fibroblasts. *J. Biol. Chem.* **259**: 8066-8069

Ochsner, M., Creba, J., Walker, J., Bentley, P., and Muakkassah-Kelly, S.F. (1990) Nafenopin, a hypolipidemic and non-genotoxic hepatocarcinogen increases intracellular calcium and transiently decreases intracellular pH in hepatocytes without generation of inositol phosphates. *Biochem. Pharmacol.* **40**: 2247-2257

Rao, M.S., and Reddy, J.K. (1987) Peroxisome proliferation and hepatocarcinogenesis. *Carcinogenesis* **8**: 631-636

Reddy, J.K., and Rao, M.S. (1986) Peroxisome proliferators and cancer: mechanisms and implications. *Trends Pharmacol. Sci.* **7**: 438-443

Scanlon, M., Williams, D.A., and Fay, F.S. (1987) A Ca^{2+}-insensitive form of Fura-2 associated with polymorphonuclear leukocytes. *J. Biol. Chem.* **262**: 6308-6312

Schiestl, R.H., and Reddy, J.K. (1990) Effect of peroxisomal proliferators on intrachromosomal and interchromosomal recombination in yeast. *Carcinogenesis* **11**: 173-176

Shuttleworth, T.J., and Thompson, J.L. (1991) Effect of temperature on receptor-activated changes in [Ca^{2+}]$_i$ and their determination using fluorescent probes. *J. Biol. Chem.* **266**: 1410-1414

Wahl, M., Lucherini, M.J., and Grunstein, E. (1990) Intracellular Ca^{2+} measurement with Indo-1 in substrate-attached cells: Advantages and special considerations. *Cell Calcium* **11**: 487-500

Whitfield, J.F., Durkin, J.P., Franks, D.J., Kleine, L.P., Raptis, L., Rixon, R.H., Sikorska, M., and Walker, P.R. (1987) Calcium, cyclic AMP and protein kinase C - partners in mitogenesis. *Cancer Metastasis Rev.* **5**: 205-250

Williams, G.M., Laspia, M.F., and Dunkel, V.C. (1982) Reliability of the hepatocyte primary culture/DNA repair test in testing of coded carcinogens and noncarcinogens. *Mutat. Res.* **97**: 359-370

Wong, S.T., Winchell, L.F., McCune, B.K., Earp, H.S., Teixido, J., Massagué, J., Herman, B., and Lee, D.C. (1989) The TGF-α precursor expressed on the cell surface binds to the EGF receptor on adjacent cells, leading to signal transduction. *Cell* **56**: 495-506

Yeldandi, A.V., Subbarao, V., Reddy, J.K., and Rao, M.S. (1989) Evaluation of liver cell proliferation during ciprofibrate-induced hepatocarcinogenesis. *Cancer Lett.* **47**: 21-27

CONFOCAL FLUORESCENCE RATIO IMAGING OF

pH IN PLANT CELLS

M.D. Fricker[1], M.R. Blatt[2], and N.S. White[3]

[1] Department of Plant Sciences, University of Oxford, South Parks Road, Oxford, OX1 3RB, UK

[2] Department of Biological Sciences, University of London, Wye College, Wye, Kent, TN25 5AH, UK

[3] Department of Zoology, University of Oxford, South Parks Road, Oxford, OX1 3PS, UK

INTRODUCTION

Fluorescent probes offer unparalleled opportunities to visualise dynamic events within living cells with a minimum of perturbation. Combined with ultra-sensitive imaging systems and powerful computer processing, quantitative measurements of ephemeral gradients, transients and oscillations are possible. Studies in plants have concentrated on measurements of cytosolic calcium ($[Ca^{2+}]_i$) and protons ($[H^+]_i$) to assess their role in signal transduction and ion homeostasis (see reviews by Hepler and Callaham, 1991; Fricker et al., 1992; Read et al., 1992). Results from stomatal guard cells indicate highly heterogeneous, spatially localised changes in $[Ca^{2+}]_i$ of varying amplitude in response to a number of stimuli (e.g. Gilroy et al., 1991; Fricker et al., 1991). In many plant systems, including guard cells, $[Ca^{2+}]_i$ and $[H^+]_i$ directly regulate membrane ion transporters at the tonoplast and plasma membrane as an integral part of a wider signalling network (Hepler and Wayne, 1985; Johannes et al., 1991; Blatt, 1992). Our knowledge of these transport mechanisms draws heavily on electrophysiological studies using patch-clamp and whole-cell voltage clamp techniques (Hedrich and Schroeder, 1989; Tester, 1990; Blatt, 1992). A major thrust of our work is to combine optical and electrophysiological approaches to allow unambiguous interpretation of signalling sequences during transduction of external stimuli. We have adopted confocal microscopy as the optimum solution for fluorescence quantitation in living tissue, and voltage-clamping in intact cells for electrophysiological analysis of membrane transporters. Voltage clamping was chosen in

Biotechnology Applications of Microinjection, Microscopic Imaging, and Fluorescence, Edited by P.H. Bach et al., Plenum Press, New York, 1993

153

preference to patch-clamping, as patching involves considerable trauma during protoplast isolation. With patching the cytoplasm is also massively diluted by the contents of the pipette, potentially disrupting signalling networks (Blatt, 1992).

Ratio Imaging and Confocal Scanning Laser Microscopy

Fluorescent ratio imaging is becoming increasingly important for visualising temporal and spatial dynamics of cytosolic ion activities. Monitoring of a shift in either the excitation or emission spectrum of the dye on binding to the ion of interest can be achieved by measurement of the fluorescence at two wavelengths, typically corresponding to the peaks for the free and bound forms. The ratio between these values automatically corrects for local variations in dye concentration, cell thickness, dye bleaching or leakage (Bright *et al.*, 1989; see also the special issue on calcium measurement in *Cell Calcium* volume **11**, 1990).

White *et al.* (1987) and Shotton (1989) have reviewed the uses of confocal scanning laser microscopy (CLSM), which permits non-invasive blur-free optical sectioning from intact, living tissues. Removal of the out-of-focus information improves the accuracy of quantitation from a defined sampling volume (voxel) in the specimen and significantly reduces the contribution of autofluorescence from structures lying outside the focal plane, such as cell walls. Optical sectioning also effectively excludes signal from dye in the barrel of the microinjection pipette used for loading, which can therefore remain in place during the experiment. This opens up the possibility of simultaneous electrophysiological measurements which do not compromise the optical techniques.

Implementation of ratio measurements on CLSM instruments requires lasers with appropriate excitation wavelengths for the fluorescent probes and rigorous attention to chromatic problems to ensure correct registration of images in x, y and z (Fricker and White, 1992). A range of dyes are now available for key ions such as Ca^{2+} and H^+ which demand precisely tailored wavelength combinations. The isobestic excitation point (pH independent wavelength) of the pH dye, 2′,7′-bis-(2-carboxyethyl)-5(and-6)carboxyfluorescein acetoxy-methylester (BCECF), conveniently coincides with the 442 nm line from HeCd lasers, whilst the pH sensitive wavelength (490-505 nm) is matched to the 488 nm line from argon ion lasers. These excitation wavelengths are also appropriate for the new long wavelength calcium dye, Fura-Red^{TM} (Molecular Probes), whose emission maximum at 600-660 nm is sufficiently well separated from that of BCECF at 530 nm to permit simultaneous monitoring of $[H^+]_i$ and $[Ca^{2+}]_i$ with dual-excitation/dual-emission CLSM. A single emission wavelength for each dye allows broad band detection and consequently good signal to noise, but measurements require correction for "bleed-through" between the two emission wavelengths for accurate quantitation. We have examined an alternative method for imaging Ca^{2+} using Indo-1 (325 nm excitation) and dual emission ratioing (405 nm and 480 nm), which offers the highest temporal resolution since both wavelengths can be collected simultaneously (Fricker and White, 1992). The introduction of a UV laser (325 nm He-Cd) into the system also offers the potential for highly controlled and spatially targeted release of caged probes permitting direct but subtle manipulation of key second messengers (e.g. Blatt *et al.*, 1990; Gilroy *et al.*, 1990).

In this paper we describe our current progress in the development of dual excitation/dual emission confocal ratio imaging with examples from the measurement of $[H^+]_i$ in stomatal guard cells using BCECF.

MATERIALS AND METHODS

Multiple-Laser Excitation CLSM

Our multiple-laser CLSM is based around a commercial instrument (Bio-Rad MRC600) attached to an inverted microscope (Nikon Diaphot). The installation is fully described in Fricker and White (1992), and only components relevant to imaging BCECF are described here. The excitation path was designed to deliver co-aligned beams from three separate lasers at 633 nm laser (3.5 mW, HeNe, Spektra Physics), 442 nm (11 mW, HeCd, Lambda Photometrics Ltd) and 514/488 nm (25 mW, argon-ion, ILT Ltd). The beams were combined with a series of dichroic reflectors and positioned using beam steering mirrors. Computer controlled electronic shutters ("Uniblitz", Vincent Associates, New York) were installed in the argon-ion and blue HeCd beams to enable rapid switching between excitation wavelengths during ratio measurements. Neutral density filters were used to balance roughly the intensity of the two beams.

Plant Material and Dye Loading

Epidermal strips from *Commelina communis* were mounted in an open perfusion system (perfusion medium comprised 50 mM KCl, 1 mM MES adjusted to pH 6.1 with KOH) under CO_2 free conditions as described previously (Gilroy *et al.*, 1991). Illumination between 400 nm and 420 nm was provided from the microscope transmission lamp via a variable interference filter. BCECF was loaded into the cytoplasm of individual guard cells by iontophoretic microinjection to an estimated concentration of 50 µM. Stomatal apertures ranged from 6-12 µm at the start of each experiment. The distribution of endoplasmic reticulum was visualised during parallel experiments after staining with $DiOC_6$ (1 µM, 10 minutes) and the vacuole imaged by staining with Acridine Orange (1 µM, 10 minutes).

Calibration

A series of *in vitro* calibration curves were measured using a variety of buffers designed to mimic conditions in the cytoplasm:-

i) the basal medium comprised 100 mM KCl, 20 mM NaCl, 1 mM $MgSO_4$, 10 mM MES, 10 mM HEPES, 1 µM BCECF adjusted with KOH to the desired pH (after Bright *et al.*, 1989);

ii) viscosity was increased by the addition of 60% sucrose (Zhang *et al.*, 1991);

iii) hydrophobicity was altered by inclusion of 25% ethanol (Russ *et al.*, 1991). Measurements were made in a chamber constructed from a multiwell microscope slide and coverslip. *In vivo* calibration was attempted using nigericin (10 µg/ml final concentration from a 10 mg/ml stock in ethanol).

Data Collection and Analysis

Cells were pre-focussed initially by confocal reflection contrast with the red 633 nm HeNe laser, to minimise photobleaching of BCECF or by conventional bright field transmission. Confocal fluorescence optical serial sections were collected using 442 nm (HeCd) and 488 nm (argon-ion) excitation and emissions at 540 ± 15 nm (BCECF) and above 600 nm (chloroplast autofluorescence). The electronic shutters switching between the lasers were synchronised with the scanning microscope by the data collection software (TCSM^{TM}, BioRad Microsciences Ltd).

Sampling rates depended on the number of image lines collected, the number of frames integrated and the disc writing time for storage. Precise quantitation of faint fluorescence requires photon counting. The MRC600 photon counting discrimination circuitry can distinguish typically 5-10 discrete levels per voxel per scan and therefore 6-10 scans per image yields around 60 independent intervals. The photon counting hardware allows precise control of black level. The TCSM software also allows sample background subtraction and calculation of the average ratio from a number of user-defined regions, such as whole cells or areas of the same cell, during each frame scanned. Graphical presentation of the ratios from these regions allowed dynamic changes to be followed during the experiment and synchronised with treatments via on-screen annotation of the traces. Post-processing and calculation of ratio images was performed using the SOM^{TM} v 4.8 and COMOS^{TM} v 6.1 programs (Bio-Rad Microsciences Ltd). After subtraction of background signals, the ratio image was calculated pixel by pixel for the 488 nm image divided by the 442 nm image. Regions less than 1/32 of the maximum 442 nm image intensity were used to construct a mask that excluded the corresponding regions of high variance, such as background, from each ratio image. Additional masking was used to exclude regions where the signal was approaching saturation and would thus yield distorted values. Final ratio images were pseudocolour, coded using one of three display look-up tables (LUT). In the first, the ratio was represented by 255 levels of fully saturated colour ranging from blue (low activity) through green to red (high activity). In the other two LUTs, the colour scale representing the ratio values was restricted to 64 levels. This scale was repeated in four bands representing four levels of dye concentration determined from the isobestic 442 nm image. In one format the bands were scaled in intensity, in the other by their level of colour saturation. With this type of coding both ratio and dye concentration data are accessible from a single image and give an indication of the confidence in the ratio value presented.

RESULTS

Figure 1 shows the results from a typical dual excitation ratio imaging experiment. BCECF was microinjected into a single guard cell of a stomatal complex within an epidermal peel from *Commelina communis*. The distribution of the dye was compared with the morphology of different sub-cellular compartments visualised directly or with appropriate vital stains. The position of the nucleus was readily visible under conventional bright field (Figure 1a). Chloroplasts were visible both in bright field images (Figure 1a) and from autofluorescence with excitation at 442 nm (Figure 1b). Dual emission detection allowed simultaneous monitoring of dye distribution and chloroplast position in 3-D (Figure 1c). Although much of the grey level intensity resolution has been sacrificed in order to reproduce the merged image in black and white, there does not appear to be any significant accumulation of BCECF in the chloroplasts.

The vacuole in guard cells accumulates Acridine Orange (Figure 1d), which can be used as a vital probe for measurement of vacuolar volume during stomatal movements (Fricker and White, 1990). The fluorescence was localised in a single compartment in greater than 80% cells examined, negatively staining the chloroplasts, nucleus and peripheral cytoplasm. In most experiments BCECF loaded by microinjection was excluded from the region typically occupied by the vacuole. Unsuccessful loading where the

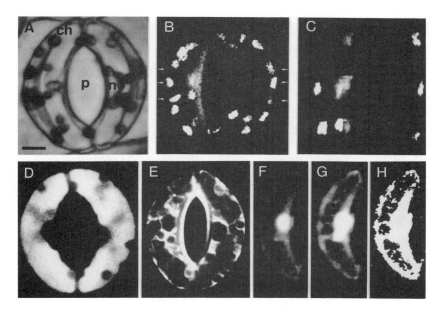

Figure 1. Comparative cellular morphology and ratio imaging of BCECF using CLSM in a stomatal complex from *Commelina communis*. **A** - Typical brightfield image of a stomatal complex collected using a transmission detector attached to the CLSM. n = nucleus; ch = chloroplast; p = pore. Scale bar = 8 μm. **B** - Merged image of chloroplast autofluorescence (bright features) and distribution of BCECF fluorescence with excitation at 442 nm (pH independent wavelength). Projection of 15 optical sections collected at 1 minute intervals simultaneously at > 600 nm (chloroplasts) and 540±15 nm (BCECF). **C** - Three *xz* sections extracted from the 3D image shown in B along the transects indicated. **D** - Guard cell vacuole morphology revealed by staining with Acridine Orange (1 μM, 10 minutes) and excitation at 488 nm. Single optical section near the mid plane of the complex shown in A. **E** - Distribution of endoplasmic reticulum and mitochondria after staining with $DiOC_6$ (1 μM, 10 minutes) and excitation at 488 nm. Maximum projection of 5 sections collected at 1 micron intervals near the midplane of the cell. **F** - Image of a guard cell after iontophoretic micro-injection of BCECF with excitation at 442 nm (pH independent wavelength) and emission at 540 ± 15nm. Average of 8 frames collected at the mid plane of the cell. Note that the injection needle near the base of the cell is not visible in this optical section. **G** - Image of the same cell as D with excitation at 488 nm (pH dependent wavelength). **H** - Ratio image (488 nm / 442 nm) with intensity scale coded such that high intensity represents high pH.

electrode penetrated the tonoplast were readily visible and rejected from further analysis. Acridine Orange was also used to follow dynamics of vacuolar morphology during stomatal movements. The number of cytoplasmic strands traversing the vacuole tended to increase with decreasing aperture. Some vesiculation occurred during very rapid closure induced by dramatic reduction of external $[K^+]$. Acridine Orange accumulated to a lesser extent in

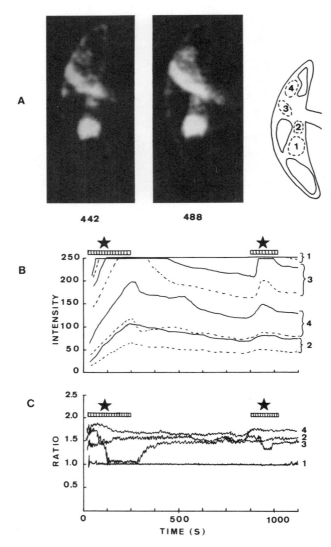

Figure 2. On-line confocal ratio measurements of BCECF microinjected into a single stomatal guard cell of *Commelina communis*. Images were collected sequentially with excitation at 442 nm (pH independent wavelength) and 488 nm (pH dependent wavelength) at a focus plane including part of the microinjection pipette near the base of the cell and several chloroplasts, using a Nikon Fluor 40 x 0.85 N.A. lens and the TSCM^{TM} (Bio-Rad Microsciences Ltd) software. **A.** The average intensity at each wavelength was measured for a number of user defined regions superimposed on the displayed images. **B.** The averaged intensity value for each wavelength after background subtraction. **C.** The calculated ratio values for these regions were graphically displayed continuously during the experiment. Traces were taken from the screen. The duration of the dye injection is indicated by the bar under the asterisk.

vacuoles of subsidiary and epidermal cells by comparison to guard cells, concentrating instead in the nucleus. Acridine Orange is a weak amine and is trapped in acidic compartments when protonated and binds to negatively charged polymers, such as DNA, RNA and mucopolysaccharides. The differential behaviour in the various cell types may be a consequence of variation both in pH gradients across the tonoplast (Penny and Bowling, 1975; Bowling and Edwards, 1984) and the bulk cytoplasmic pH.

The potential sensitive dye, $DiOC_6$, is one of a family of carbocyanine dyes that preferentially accumulate in membranes with a net charge such as the endoplasmic reticulum. In guard cells fluorescence from $DiOC_6$ occurred in trans-vacuolar strands, discrete punctate dots around the cell periphery (mitochondria or possibly strands of endoplasmic reticulum viewed in transverse section), and larger aggregations around the nucleus (Figure 1e). The precise balance between reticulate and punctate staining was dependent on the dye concentration, incubation time and additional, but undefined factors, related to the state of the tissue. Typically, mitochondria were visible during shorter incubations and at lower concentrations than required for endoplasmic reticulum. Much of this detail was obscured with conventional epifluorescence observation as carbocyanine dyes appeared to bind to certain wall components (probably in the cuticle) and fluoresced strongly.

BCECF fluorescence was visible on excitation at 442 nm (Figure 1f) and 488 nm (Figure 1g) in the peripheral cytoplasm and also in cytoplasmic strands traversing the vacuole. Within a single confocal optical section (approximately 1 μm thick) the signal was heterogenous. The high intensity throughout the nucleus of both wavelengths indicated the presence of substantial concentrations of dye within this organelle. The ratio image (Figure 1h) indicated that pH remained relatively constant within the bulk cytoplasm despite the variation in absolute intensity in different regions. This emphasises the importance of the ratio approach over single wavelength measurements in correcting for variation in dye concentration. The apparent lower pH in the nucleus was an artifact due to saturation of the 488 nm signal necessary to detect the cytoplasmic details. Additional masking was normally used to exclude these artifactual regions (see Figure 3 a-c). Further information on the dye concentration was obtained from the dual colour/intensity LUT or colour/saturation LUT representations, but cannot be reproduced here in black and white.

Calculation and display of ratio images is usually too slow without special hardware for updating during the experiment. However, one approach that allows continuous monitoring during the course of the experiment is to follow ratios spatially averaged over a number of user defined regions (Figure 2a). The averaged intensity values (Figure 2b) and the corresponding ratios (Figure 2c) can be displayed in a continuous graph, providing a recorded history of the experiment.

Ratio images are usually calculated post-capture and pseudocolour coded. A gallery of pseudocolour coded ratio *xy* images taken at timed intervals gives some indication of pH homeostasis and spatial heterogeneity within the cytoplasm (Figure 3d). Alternatively changes in a single line of voxels across the image portrayed over time accentuate temporal variation (Figure 3e). Such "*xt*" images may also be collected by repeated scanning of a single line giving temporal resolution as high as 2 milliseconds. Extraction of a row from this image gives a confocal line trace of ion activity against time for a single voxel (Figure 3f).

During microinjection, the local pH appeared to increase near the electrode (Figure 3 d, e, f). Recovery was rapid following injection, but the increase in pH was repeated after subsequent injections. Within the cytoplasm there appeared to be both regions of higher pH

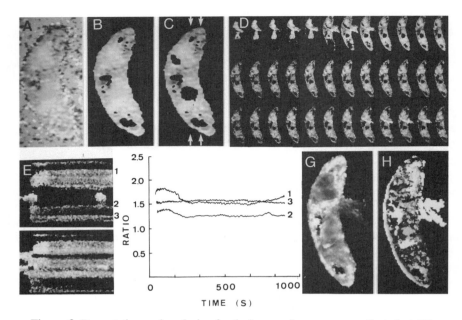

Figure 3. Presentation and analysis of ratio images. Images were collected at 442 nm and 488 nm near the base of a guard cell during iontophoretic microinjection of BCECF (see Fig 2). **A** - Unmasked ratio image. High intensities correspond to high pH. **B** - Masked at 1/32 maximum to exclude regions of low signal. **C** - The same image as B with additional masking to exclude regions approaching saturation at either wavelength (predominantly the nucleus and near the tip of the injection pipette). **D** - Gallery of ratio images over time. 36 views from a series of 114 images collected at 10 second intervals during the microinjection sequence. Dye was injected at the start of the experiment and at the points marked with an asterisk. **E** - Presentation of *xt* images to visualise changes in pH over time for the line transects indicated on image B. **F** - Plot of ratio against time for three rows of pixels indicated in E. **G** - Projection of the minimum $[H^+]$ recorded from each voxel during the experiment shown in D. Note that the intensity scale has been recoded to reflect $[H^+]$ rather than pH. High intensities correspond to high $[H^+]$. **H** - Time at which these values were recorded. White early in the experiment, black late.

around the chloroplasts and highly localised, though smaller (about 1 µm) regions of low pH which were sustained over time. The combined intensity and ratio images indicated that these regions had ample dye concentration for reliable measurements. Calibration of the *in vivo* ratio values was not very reproducible as the dye leaked out of ionophore clamped cells (see also Pheasant and Hepler, 1987; Dixon *et al.*, 1989). Relating *in vivo* ratios to *in vitro* calibration curves depends on reliable models of the cytoplasm composition. Addition of sucrose or ethanol to the calibration solution altered the dye response, but is more typical of the *in vivo* environment. With this calibration, the ratio scale ranged from about pH 8.0 (white) to pH 6.5 (black). Typical resting pH values were pH 7.4 in the bulk cytoplasm increasing to pH 7.8 near the chloroplasts. The smaller regions of low pH were estimated to be pH 6.8 or lower.

We have developed tools to extract a subset of information from the complete image series based on a set of user-defined rules appropriate to the biological question under investigation. One approach is to compress or reduce the data along one axis by projection. A typical example is shown in Figure 3 g and h. The projection algorithm has extracted the minimum $[H^+]$ recorded for each position in the xy image over time and displayed this value as a single composite image. (Note that the intensity scale reflects $[H^+]$ rather than pH in this image - low $[H^+]$ is thus coded as dark). In addition, the time when this value or peak response was reached is also found. This temporal image is displayed in the second panel using a co-ordinate or position projection coded with an intensity scale of white (early in the sequence) to black (later). This is the temporal equivalent of a HEIGHT projection (White et al., 1991). The $[H^+]$ image confirms the localisation of low $[H^+]$ around the chloroplasts and the electrode. The predominance of high intensities in the "time" image near the electrode indicate these values were recorded early in the sequence, whilst the heterogeneity in the lower part of the cell indicates there was no major trend in the timing of pH changes over this region. The strong left hand edge of the cell in the "time" image suggests that the stoma has closed slightly during the course of the experiment.

DISCUSSION

The ratio recorded from single, unstimulated guard cells was heterogeneous, though at this stage the absolute pH values must be regarded with caution until we have more rigorously tested the reliability of our *in vivo* calibrations. It is not clear whether the increase in pH near the electrode during injection was a direct consequence of the current applied driving protons into the pipette (or OH^- ions into the cytoplasm) or perhaps represented temporary and local breakdown of the normal $[H^+]_i$ homeostasis within the cell. The higher pH around the chloroplasts was independent of the injection period and was sustained over time. Two possible explanations exist:-

1. There may be genuine proton gradients set up by

    a) chloroplast electron transport,

    b) proton imbalance through the action of the various translocators, and

    c) alkalinisation from CO_2 fixation.

All these factors would be expected to increase pH in the surrounding cytoplasm, though CO_2 fixation by guard cell chloroplasts remains a controversial area (e.g. Shimazaki, 1989).

2. The two excitation wavelengths may be differentially absorbed by chloroplast pigments and therefore alter the level of dye excitation. The predicted preferential absorption of the 442 nm excitation wavelength by chlorophyll would yield apparently higher pH around the chloroplasts. These features would still occur with conventional imaging, but would normally be hidden in the blur of out-of-focus information. We are currently examining means to distinguish between these possibilities, including inhibitors of chloroplast proton transport and mapping the distribution of pH insensitive dyes with excitation at 442 nm and 488 nm.

The smaller regions of lower pH may result from compartmentalisation of BCECF into more acidic vesicles such as the endoplasmic reticulum. Our initial attempts to correlate distribution of BCECF with endoplasmic reticulum have been indirect. It may be possible to perform more definitive co-localisation using double labelling with potential sensitive dyes other than $DiOC_6$ whose spectra do not overlap with BCECF.

Reproduction of pseudocolour ratio images is expensive and it is often difficult for the reader to assimilate information from a series of separate images. This is compounded when several parameters are measured simultaneously and the dimensionality of the image data increases e.g. the 3-D measurement of several ions over time produces data in 5 dimensions. It is becoming increasingly important, therefore, to devise means of extracting relevant data and presenting it in a concise form. In this example, a selected temporal sub-set of the 114 original images has been condensed by projection and displayed in two images. The minimum $[H^+]$ values recorded may be related to whether particular pH-dependent ion channels are likely to have gated open. Using our latest projection routines for limited processing over a local or segmented region we can extract and code the time interval over which ion activities were elevated above a particular threshold, for example, or when the frequency of oscillations in ion activity became significant. A gallery or sequence of such images allows comparison of selected parameters imaged over a series of experimental treatments in a single figure.

CONCLUSIONS

The CLSM system described here is useful for dual excitation measurements of pH using the pH sensitive dye BCECF with significantly improved quantitation and 3-D localisation in comparison to photometry or camera based imaging systems. The spatial resolution of the CLSM can be traded against the amount of temporal resolution required. Single points in 3-D could be confocally sampled at microsecond intervals or line scans repeated at 2 millisecond steps to follow a number of spatially well defined voxels over time. 2-D images typically require 0.25 to several seconds, whilst complete 3-D data sets requiring a few minutes are more appropriate for examining dye distribution at the highest resolution or for slowly changing systems such as standing $[H^+]_i$ gradients or gentle oscillations in $[H^+]_i$ through the cell cycle. Powerful digital techniques are also emerging to extract and display quantitative information from multi-dimensional images using interactive selection criteria and projection algorithms. Such protocols are likely to prove essential to realise the full potential of quantitative microscopy in practice.

ACKNOWLEDGEMENTS

NSW thanks SERC for financial support. Equipment for this work was obtained on grants from the Nuffield Foundation, Royal Society and AFRC to MDF; and on SERC grants to NSW and MDF, and to D.M. Shotton, Zoology Department, Oxford. We also thank Bio-Rad Microsciences Ltd, Spindler and Hoyer Ltd. and Lambda Photometrics for technical information and the loan of equipment.

REFERENCES

Blatt, M.R. (1992) Ion channel gating in plants: physiological implications and integration for stomatal function. *J. Membr. Biol.* (in press)

Blatt, M.R., Thiel, G., and Trentham, D.R. (1990) Reversible inactivation of K^+ channels of *Vicia* stomatal guard cells following the photolysis of caged inositol 1,4,5-triphosphate. *Nature* **346**: 766-769

Bright, G.R., Fisher, G.W., Rogowska, J., and Taylor, D.L. (1989) Fluorescence ratio imaging microscopy. *Meths. Cell Biol.* **30**: 157

Bowling, D.J.F., and Edwards, A. (1984) pH gradients in the stomatal complex of *Tradescantia virginiana*. *J. Exp. Bot.* **35**: 1641-1645

Callaham, D.A., and Hepler, P.K. (1991) Measurement of free calcium in plant cells, *in:* "Cellular Calcium: A Practical Approach," J.G. McCormack and P.H. Cobbold, eds., IRL Press, Oxford, 383-410

Dixon, G.K., Brownlee, C., and Merrett, M.J. (1989) Measurement of internal pH in the coccolithophore *Emilianai huxleyi* using 2',7'-bis-(2-carboxyethyl)-5(and-6)carboxyfluorescein acetoxymethylester and digital imaging microscopy. *Planta* **178**: 443-449

Fricker, M.D., Gilroy, S., Read, N.D., and Trewavas, A.J. (1991) Visualisation and measurement of the calcium message in guard cells, *in:* "Molecular Biology of Plant Development," G.I. Jenkins and W. Schuch, eds., *SEB Symp. Series* **44**: 177-190

Fricker, M.D., Tester, M., and Gilroy, S.G. (1992) Fluorescent and luminescent techniques to probe ion activities in living plant cells, *in:* "Fluorescent Probes for Biological Activity of Living Cells - A Practical Guide," W.T. Mason and G. Relf, eds., Academic Press, London (in press)

Fricker, M.D., and White, N.S. (1990) Volume measurements of gurd cell vacuoles during stomatal movements using confocal microscopy. *Trans. Royal Microscopical Soc.* **1**: 345-348

Fricker, M.D., and White, N.S. (1992) Wavelength considerations in confocal microscopy of botanical specimens. *J. Microscopy* **166**: 29-42

Gilroy, S.G., Fricker, M.D., Read, N.D., and Trewavas, A.J. (1991) Role of calcium in signal transduction of *Commelina* guard cells. *Plant Cell* **3**: 333-344

Gilroy, S.G., Read, N.D., and Trewavas, A.J. (1990) Elevation of cytoplasmic calcium by caged calcium or caged inositol triphosphate initiates stomatal closure. *Nature* **346**: 769-771

Hedrich, R., and Schroeder, J.I. (1989) The physiology of ion channels and electrogenic pumps in higher plants. *Ann. Rev. Plant Physiol.* **40**: 539-569

Hepler, P.K., and Wayne, R.O. (1985) Calcium and plant development. *Ann. Rev. Plant Physiol.* **36**: 397-439

Johannes, E., Brosnan, J.M., and Sanders, D. (1991) Calcium channels and signal transduction in plant cells. *Bioessays* **13**: 331-336

Penny, M.G., and Bowling, D.J.F. (1975) Direct determination of pH in the stomatal complex of *Commelina*. *Planta* **122**: 209-212

Pheasant, D.J., and Hepler, P.K. (1987) Intracellular pH change is not detectable during the metaphase/anaphase transition in *Tradescantia* stamen hair cells. *Eur. J. Cell Biol.* **43**: 10-13

Read, N.D., Allan, W.T.G., Knight, H., Knight, M.R., Malho, R., Russell, A., Shacklock, P.S., and Trewavas, A.J. (1992) Imaging and measurement of cytosolic free calcium in plant and fungal cells. *J. Microscopy* **166**: 57-86

Russ, U., Grolig, F., and Wagner, G. (1991) Changes in cytoplasmic free Ca^{2+} in the green alga *Mougeotia scalaris* as monitored with Indo-1, and their effect on the velocity of chloroplast movements. *Planta* **184**: 105-112

Shimazaki, K.-I. (1989) Ribulosebisphosphate carboxylase activity and photosynthetic evolution rate in *Vicia* guard-cell protoplasts. *Plant Physiol.* **91**: 459-463

Shotton, D.M. (1989) Review: confocal scanning optical microscopy and its applications for biological specimens. *J.Cell Science* **94**: 175-206

Tester, M. (1990) Plant ion channels: whole-cell and single-channel studies. *New Phytol.* **114**: 305-340

White, J.G., Amos, W.B., and Fordham, M. (1987) An evaluation of confocal versus conventional imaging of biological structures by fluorescence light microscopy. *J. Cell Biol.* **105**: 41-48

White, N.S., Fricker, M.D., and Shotton, D.M. (1991) Quantitative visualisation of 3D biological CLSM images. *Scanning* **13**: 51-53

Zhang, D.H., Callaham, D.A., and Hepler, P.K. (1990) Regulation of anaphase chromosome motion in *Tradescantia* stamen hair cells by calcium and related signalling agents. *J. Cell Biol.* **111**: 171-182

ANALYSES OF THE CALCIUM AND PHOSPHORUS DISTRIBUTION IN PREDENTINE TISSUE OF RAT INCISORS

H.P. Wiesmann,[1] U. Plate,[2] H.J. Höhling[1], and R.H. Barckhaus[1]

[1]Institut für Medizinische Physik und Biophysik, Schmeddingstraße 50, 4400 Münster, Germany
[2]Physikalisches Institut, Wilhelm Klemm Straße 10, 4400 Münster, Germany

INTRODUCTION

The rat incisor is a continuously growing tooth and so the physiological biomineralization of enamel, dentine, and cementum continues throughout life. In dentinogenesis, the odontoblasts form an extracelluar matrix, the predentine consisting of collagen, phosphoproteins, proteoglycans and other matrix macromolecules. After a thin mineral sheet has been developed, matrix vesicles are no longer observable and predentine mineralizes continuously at a distance up to 20 μm away from the odontoblasts directly at the dentine border. In contrast to the mineralization of bones, in dentine no remodelling processes occurs. Thus dentine is an excellent model to study the process of hard tissue formation in a collagen rich system.

The apatite mineral in dentine consists mainly of calcium and phosphate so it is important to analyse their distribution in predentine, especially at the mineralization front, i.e. the predentine - dentine border. Quantitative electron probe microanalysis in predentine was first reported by Höhling et al. (1967) and continued by Ashton et al. (1973), Nicholson et al. (1977) and Höhling et al. (1991). In these studies cryosections with a thickness range of 4-8 μm and dry semithin sections of shockfrozen, freeze-dried embedded tissue have been analysed. However, cryosections of this thickness reveal no histological details of the predentine. Therefore it is difficult to relate the analyzed areas to the morphology. Further, they allow only a retricted lateral resolution concerning elemental

Biotechnology Applications of Microinjection, Microscopic Imaging, and Fluorescence, Edited by P.H. Bach *et al.*, Plenum Press, New York, 1993

165

analysis especially near the dentine border. In semithin dry resin sections the resin may cause a dislocation of some elements in the tissue during the infiltration process.

In this study we present for the first time analyses on ultrathin cryosections of predentine. Quantitative results for the analyses of calcium and phosphorus were received by the energy dispersive X-ray microanalyses (EDX) and they were complemented by qualitative electron energy loss spectra (EELS). The qualitative EELS analyses were carried out to get information about the calcium distribution with a high lateral resolution even at the mineralization front.

MATERIALS AND METHODS

Incisors of young Wistar rats (weight about 150 g) were prepared and cryofixed in liquid nitrogen-cooled propane as described earlier by Plate *et al.* (1992). Tissue preparation and cryofixation took less than three minutes.

Cryosections with a thickness of 80 nm to 100 nm were cut with a Reichert FC4-Ultracut cryo-ultramicrotome and transferred from the glass knife onto a Pioloform- and carbon-coated 75-mesh copper grid by using an eye-lash; further, cryosections were transferred to a 300-mesh copper grid. The ultrathin sections were then freeze dried. The preparation of the cryosections was described in detail by Zierold (1991).

The cryosections were studied in a Philips transmission microscope 301 and in a Zeiss EM 902. The Philips 301 was equipped with a Si(Li)-detector and an EDAX 9100 system which were used to get quantitative results. All quantitative results are given in weight-percent per dry mass (g/100 g) and the Ca/P-ratios are given as weight ratios. EELS spectra were taken with the Zeiss EM 902 by a scintillator photomultiplier combination.

RESULTS

Figure 1 shows the results of the EDX microanalyses for calcium, phosphorus, and the Ca/P-ratios in the odontoblast region and in predentine. The concentrations are plotted against the distance of the analyzed areas (1 x 1 μm) from the mineralization front, i.e. the predentine/dentine border. The extracellular predentine had a width range of about 10-12 μm. Therefore the element concentrations at higher distances than 12 μm were measured in the odontoblast region. The calcium concentration is in the range of about 0.3% and at 1 μm from the dentine border shows a slight increase. Phosphorus concentrations in the odontoblasts rise up to 2.0% and in the predentine are in the range of 0.2%-0.3%. Also at a distance of 1 μm from the mineralisation front a slight increase is visible. In predentine the Ca/P-ratio is less than 1 and it starts rising at a distance of about 3 μm in front of the mineralization front.

EELS spectra were taken at distances of about 2 μm and 100 nm from the mineralisation front and also at the primary dentine crystallites which can be described as chains of dots and as needles (Figure 2). Figure 3 shows only a small increase of the Ca-peak in the spectrum being obtained at a distance of 100 nm (B) compared to the spectrum at a distance of 2 μm (A). The Ca-peak rises up in the spectrum which is taken from the first dentine crystallites (C).

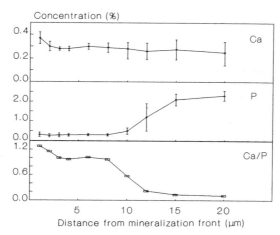

Figure 1. EDX microanalysis of Ca and P concentrations and the Ca/P ratio in ultrathin cryosections of predentine plotted against the distance of the scanning area from the mineralisation front. Concentrations are given in percentage / dry mass ± SD, number of measurements at each point is more than 25. Note: predentine is between the mineralisation front (0 μm) and the odontoblasts layer which appears at a distance from the odontoblasts of about 10-12 μm.

DISCUSSION

A major problem with the analyses of the element distribution in predentine is the preparation of the dentine tissue. Factors in the preparation steps which may affect the element distribution of the tissue are shown in Table 1 for resin sections and cryosections.

The incisors were cryofixed as fast as possible in order to prevent postmortal element diffusion. Also dislocations of the element distribution caused by ice crystal damage have to be taken into account. Consequently a scanning area of 1 x 1 μm, which is larger than the largest ice crystals, was used for analysis. An important disadvantage of resin sections is that during the infiltration process resin may dramatically displace elements. Therefore ultrathin cryosections seem to be the optimal way of preparing specimens for elemental analyses.

Earlier investigations of thick cryosections resulted in element concentrations in the range of 0.2-0.4% for Ca and 1-4% for P (Höhling et al. 1968), 0.1-0.3% (mean 0.28%) for Ca and 0.3-3.4% for P (Ashton et al. 1971) and up to 4% for Ca in enriched zones (Nicholson et al. 1977). So generally our results for calcium are in quite good agreement with these earlier results. The previously reported phosphorus concentrations of the 4-8 μm cryosections are higher than the phosphorus concentrations found in our ultrathin cryosections. Using cryosections in a thickness range of about 4-8 μm it is impossible to distinguish between odontoblast processes and extracellular predentine. Phosphorus concentrations measured in predentine of 4-8 μm cryosections are generally "mixed" values of extracellular predentine and odontoblast processes. The phosphorus content in the odontoblast processes is as high as in the odontoblast bodies which have phosphorus concentrations up to 2.0%. So the phosphorus content in the predentine region seems to be higher in 4-8 μm cryosections than in the predentine of ultrathin cryosections in which it is easy to visualize the odontoblast processes.

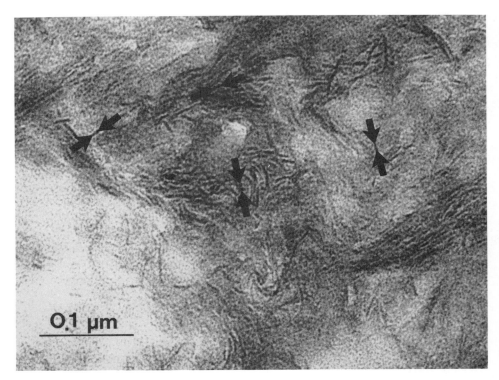

Figure 2. Earliest crystal formations can be described as chains of dots (arrows designate the dot to dot distances) along the collagen-microfibrils which rapidly coalesce to form parallel arranged "needlelike" crystallites.

The EDX analyses give an indication of an increase of calcium and phosphorus concentrations near the mineralization front. However, the lateral resolution of EDX analyses concerning calcium and phosphorus near the dentine border is limited by the secondary excitation of the calcium and phosphorus atoms in the dentine. The secondary excitation was discussed e.g., by Felsmann (1987) and Höhling *et al.* (1991).

By contrast EELS analysis has a much higher lateral resolution than EDX analysis. For example, Weiss and Carpenter (1992) showed that in a STEM the lateral resolution differs from the diameter of the electron beam only in a submicrometer range. The sensitivity of EELS element analysis is good for calcium but phosphorus analysis presents problems. In the EELS spectra (Figure 3) only the calcium peaks are visible. The EELS analyses show a slight increase of the calcium peak in the predentine directly beside the primary mineralisation (chains of dots, Figure 2) appears. Thus EELS analyses confirm the results of the EDX microanalyses.

In our analyses, especially on the ultrathin cryosections, the results of Höhling *et al.* (1991) have been supported that a zone of Ca-enrichment exists at the mineralization front in the predentine. However, the zone was found to be smaller (about 1 μm wide) than

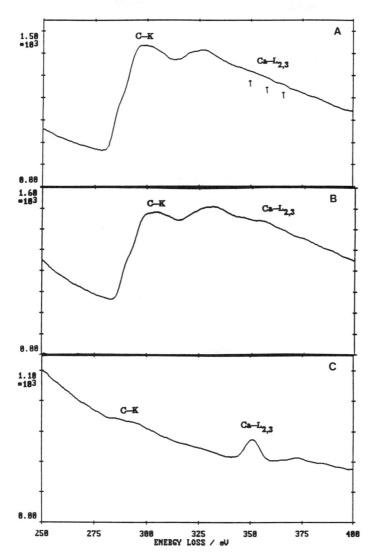

Figure 3. EELS spectra in predentine of ultrathin cryo-sectioned rat incisors (thickness range of about 80-90 nm), **A)** in the middle of the predentine, **B)** 100 nm in front of the mineralisation front, and **C)** directly taken from the primary crystallites (chains of dots)

Table 1. Problems of tissue preparation for elemental analysis

PREPARATION STEP	CRYOSECTION	RESIN SECTION
preparation of tooth	postmortal element diffusion	postmortal element diffusion
cryofixation	ice crystal damage[1]	ice crystal damage[1]
cryosectioning	mechanical deformation and micromelting ?[2]	
freeze-drying	section: shrinkage (10-20%)[3]	tissue: shrinkage possible
vacuum embedding process		dramatically element displacements possible
dry-sectioning of embedded tissue		mechanical deformation possible
exposure of sections to air	slight changes of element concentrations and in morphology[4]	uncritical
storage	even under vacuum problematical[4]	uncritical

[1] Echlin (1991)
[2] Kirck et al. (1991) and Richter et al. (1991)
[3] Zierold (1984)
[4] Zglinicki and Zierold (1988)

assumed by Höhling *et al.* (1991). It is conceivable that this small zone is a "turnover zone" in which the conditions for Ca-phosphate nucleation are prepared. Glimcher (1989) and Linde *et al.* (1989) have demonstrated that immobilized phosphoproteins can stimulate the apatite mineralization. Fujisawa (1987) showed that soluble phosphoproteins inhibit precipitation of calcium phosphate *in vitro*. Thus the calcium and soluble phosphoproteins in this "turnover zone" might be in the process of being bound to the matrix macromolecules, which induce crystal nucleation.

ACKNOWLEDGEMENTS

This work was supported by the DFG and the DGZMK.

REFERENCES

Ashton, B.A., Höhling, H.J., Nicholson, W.A.P., Zessack, U., Kriz, W., and Boyde, A. (1973) Quantitative analysis of Ca, P and S in mineralizing tissues. *Naturwissenschaften* **60**: 392-393

Echlin, P. (1991) Ice crystal damage and radiation effects in relation to microscopy and analysis at low temperatures. *J. Microsc.* **161**: 159-170

Felsmann, M. (1987) Elektronenstrahlmikroanalyse dünner Proben: Räumliche Auflösung und Untersuchung des Calzium-Transportes in der Wachstumsfuge mit Hilfe von Stronzium als Markierungselement; (Electron probe microanalysis of thin specimens: lateral resolution and investigation of the calcium transport in the epiphysial growth plate using stroncium as tracer for calcium), Thesis, Dept. of Physik, University of Münster, 26-95

Fujisawa, R., Kuboki, Y., and Sasaki, S. (1987) Effects of dentine phosphoryn on precipitation of calcium phosphate in gel *in vitro*. *Calc. Tiss. Int.* **41**: 44-47

Glimcher, M.J. (1989) Mechanism of calcification: role of collagen and collagen-phosphoprotein complexes *in vitro* and *in vivo*. *Anat. Rec.* **224**: 139-153

Höhling, H.J., Hall, T.A., and Boyde, A. (1967) Electron probe X-ray microanalysis of mineralisation in rat incisor peripheral dentine. *Naturwissenschaften* **54**: 617-618

Höhling, H.J., Hall, T.A., Boyde, A., and von Rosenstiel, A.T. (1968) Combined electron probe and electron diffraction analysis of prestages and early stages of dentine formation in rat incisor. *Calcif. Tissue Res.* **2**: 5-6

Höhling, H.J., Mishima, H., Kozawa, Y., Daimon, T., Barckhaus, R.H., and Richter, K.-D. (1991) Microprobe analyses of the potassium-calcium distribution relationship in predentine. *Scanning Microsc.* **5**: 247-255

Kirk, G., Knoff, L., and Lee, P. (1991) Surfaces of cryosections: is cryosectioning 'cutting or fracturing'?. *J. Microsc.* **161**: 445-453

Linde, A., Lussi, A., and Crenshaw, M.A. (1989) Mineral induction by immobilized polyanionic proteins. *Calc. Tiss. Int.* **44**: 286-295

Nicholson, W.A.P., Ashton, B.A., Höhling, H.J., Quint, P., Schreiber, J., Ashton, I.K., and Boyde, A. (1977) Electron Microprobe Investigations into the Process of Hard Tissue Formation. *Cell Tiss. Res.* **177**: 331-345

Plate, U., Reimer, L., Höhling, H.J., Barckhaus, R.H., Wienecke, R., Wiesmann, H.P., and Boyde, A. (1992) Analysis of the Ca-distribution in predentine by EELS and the early crystal formations in dentine by ESI and ESD. *J. Microsc.* (in press)

Richter, K., Gnägi, H., and Dubochet, J. (1991) A model for cryosectioning based on the morphology of vitrified ultrathin sections. *J. Microsc.* **163**: 19-28

von Zglinicki, T., and Zierold, K. (1989) Elemental concentrations in air-exposed and vacuum-stored cryosections of rat liver cells. *J. Microsc.* **154**: 227-235

Weiss, J.K., and Carpenter, R.W. (1992) Factors limiting the spatial resolution and sensitivity of EELS microanalysis in a STEM. *Ultramicroscopy* **40**: 339-351

Zierold, K. (1984) The morphology of ultrathin cryosections. *Ultramicroscopy* **14**: 201-210

Zierold, K. (1991) Cryofixation methods for ion localisation in cells by electron probe microanalysis: a review. *J. Microsc.* **161**: 357-366

QUANTITATIVE LOCALISATION OF HIV
PROTEINS IN MAMMALIAN CELLS

Lorene E.A. Amet[1], Michael R.H. White[1,2], Julia A. Sutton[1], Linda J. Capsey[1,3], Martin Braddock[1], Alan J. Kingsman[1,4], and Susan M. Kingsman[1]

[1]Virus Molecular Biology Group, Department of Biochemistry, South Parks Rd, Oxford OX1 3QU, UK

[2]Amersham International plc, White Lion Rd, Little Chalfont, Bucks HP7 9LL, UK

[3]Pharmacia UK Ltd, Davy Avenue, Knowlhill, Milton Keynes MK5 8PH, UK

[4]British Biotechnology Ltd, Watlington Rd, Oxford OX4 5LY, UK

INTRODUCTION

The human immunodeficiency virus (HIV-1) has the typical genetic organisation of retroviruses, with three major genes, namely *gag*, *pol* and *env*. In addition, the virus encodes several regulatory proteins. The regulation of gene expression in HIV-1 is critically dependent on the virally encoded protein (Dayton *et al.*, 1986; Fisher *et al.*, 1986). The TAT protein is essential for viral replication. TAT is a positive feedback transactivator that increases the level of gene expression from the HIV-1 long terminal repeat (LTR) and, therefore, increases the rate of its own synthesis and the synthesis of all viral proteins. The exact mechanism of TAT action has been the subject of much controversy and to date is not fully understood. The TAT transactivator protein interacts with a cis-acting element called TAR which is located immediately downstream of the transcription start-site and is, therefore, present in the 5′ untranslated regions of all HIV mRNAs (Rosen *et al.*, 1985).

Biotechnology Applications of Microinjection, Microscopic Imaging, and Fluorescence, Edited by P.H. Bach *et al.*, Plenum Press, New York, 1993

TAR RNA forms a stable bulge stem-loop structure (Muesing *et al.*, 1987) and *in vitro* studies have shown that the bulge is necessary for TAT binding (Roy *et al.*, 1990). TAT acts predominantly at the level of transcription in mammalian cells (Rice and Matthews, 1988; Jakobovits *et al.*, 1988; Jaeng, 1988). However, some studies have suggested that TAT may function post-transcriptionally to increase the translational capacity of TAR RNAs (Rosen *et al.*, 1986; Muesing *et al,*. 1987). In the *Xenopus* oocyte model system, we have shown that the major role of TAT is post-transcriptional but that this requires the oocyte nucleus for activity (Braddock *et al,* 1989; Braddock *et al.*, 1990; Braddock *et al.*, 1991; White *et al.*, this volume). The TAT protein has previously been shown to be located in the nucleus of mammalian cells and appears to be especially concentrated in the nucleolus (Hauber *et al.*, 1987; Ruben *et al.*, 1989). A nuclear/nucleolar localization signal (NOS) was previously defined and corresponds to a small cluster of basic residues (Siomi *et al.*, 1990). This domain is also involved in the binding to TAR RNAs (Dingwall *et al.*, 1989; Dingwall *et al.*, 1990; Kamine *et al.*, 1991). Mutations that alter the TAT basic domain destroy both activity and localisation. It is not clear, however, whether the loss of activity is due to defective cellular targeting or to the inability of the protein to interact with TAR RNAs. We previously dissected the two functions of the TAT basic sequence by injecting a truncated TAT48 protein, which lacked the TAT basic sequence, into the nucleus of *Xenopus* oocytes (Braddock *et al.*, 1989). This provided an excellent system to test the role of the different cellular compartments in TAT transactivation. Interestingly, we found that TAT48 was as active as the wild type TAT protein when injected into the nucleus. This suggested that, in the *Xenopus* oocyte system, of the two functions assigned to the TAT basic sequence, only the nuclear/nucleolar targeting function and not the previously described RNA binding function (Dingwall *et al.*, 1989; Dingwall et al. 1990; Kamine *et al.*, 1991) is required for transactivation. The aim of the present work has been to apply the techniques of quantitative immunofluorescence imaging to analyse the subcellular localisation of modified TAT proteins in mammalian cells and to attempt to correlate the sub-cellular localisation with TAT activation. Quantitative immunofluorescence localisation of proteins was performed using a Silicon Intensified Target (SIT) camera and an Argus 100 Image Processor supplied by Hamamatsu Photonics. A principle question of the study was whether we could separate the role of the basic patch in determining localisation and RNA binding functions of TAT protein in mammalian cells.

In order to investigate this, we constructed a gene encoding a modified TAT48 protein which could be targeted into the nucleus, but which could not interact with TAR via the bulge-binding interaction described by *in vitro* studies (Dingwall *et al.*, 1989; Dingwall *et al.*, 1990; Kamine *et al.,* 1991). For this, we fused the basic sequence for the oncogene C-Myc which is known to be a nuclear targeting signal (Dang and Lee, 1988), to the N-terminus of TAT48. TAT, TAT48 and C-MycTAT48 were expressed in COS cells and analysed both for their transactivation activities and for the subcellular localisation of the proteins. The transactivation activity of these proteins was assayed by co-transfection experiments in which the TAT protein expression vectors were co-transfected into mammalian cells together with a reporter vector containing the chloramphenicol acetyl transferase (CAT) gene under the control of the HIV-1 LTR sequence (Adams *et al.*, 1988). We report that in contrast with the *Xenopus* oocyte system, the nuclear localization of TAT48 is not sufficient for transactivation and this suggests that the TAT-TAR interaction mediated by the TAT basic sequence is required for activity in mammalian cells. This report also discusses the problems and advantages of quantitative imaging analysis of protein localization in mammalian cells.

MATERIALS AND METHODS

Construction of Tat48 and C-MycTat48 Expressing Vectors

TAT48 and C-MycTAT48 coding sequences were cloned into an SV40 origin-based plasmid (pKV 939), under the control of the immediate early human cytomegalovirus (hCMV) promoter (Figure 1). Construction of the TAT48 coding sequence was previously described (Braddock *et al.*, 1989). For the C-MycTAT48 construction, two oligonucleotides encoding the M1 nuclear localization signal of human C-Myc (pro320 to leu327), flanked by a *Bam HI* site and a *Bgl II* site were synthesised (Oxford Enzyme Group Oligo nucleotide service), purified (Maniatis *et al.*, 1990) and annealed. The resultant double-stranded oligonucleotide was ligated upstream of the TAT48 sequence.

Maintenance of Mammalian Cell Lines

HeLa and COS-7 cells were maintained in MEM and DMEM, respectively (Gibco). Media were supplemented by 10% foetal calf serum (Gibco), 2 mM glutamine, 1 x non-essential amino-acids (Gibco), and 50 U/ml of penicillin and streptomycin. The cells were grown at $37^{\circ}C$ in an atmosphere of 5% CO_2.

Transfection of Mammalian Cell Lines

One day before transfection, cells were plated at a density of 1×10^4 cells/cm^2 onto glass coverslips for immunofluorescence analysis, or in 6-cm plates for CAT activity assays. COS-7 cells were transfected using the DEAE-dextran procedure and HeLa cells by the calcium phosphate coprecipitation method (Gorman, 1982).

Assays of CAT Activity

COS-7 cells were transfected with 1 µg of transactivator coding plasmids (TAT, TAT48 or C-MycTAT48) together with 3 µg of an HIV-1 LTR CAT plasmid (pOGS210) and 1 µg of a ß-galactosidase expressing plasmid (pPE346) as an internal control. Cell extracts were made 48 hour after transfection, by sonication, and assayed for CAT activity (Gorman, 1982). CAT activities were corrected for the ß-galactosidase control. Typically, CAT assays were incubated for 1-2 hours with about 10 µg cell protein and the thin layer chromatography plates analysed using a Molecular Dynamics Phosphorimager™.

Cell Fixation and Antibody Staining

Transfected cells were grown for 48 hours before fixing in 3% paraformaldehyde. They were washed twice in 10 mM glycine/PBS for 5 minutes each, permeabilized in 1% Triton-X100/PBS for 5 minutes and washed again in 25 mM glycine/PBS for 30 minutes. Prior to addition of the primary antibody, the cells were blocked in a solution containing 1% BSA, 0.5M NaCl, 0.05% Tween-20 and 3% whole horse serum, for a minimum of 1 hour. A mouse monoclonal anti-TAT antibody (MRC, AIDS directed programme, reagents catalogue) was added at a final dilution of 1/500 in blocking solution and was left on the cells for 10-15 hours at 4°C with gentle shaking. Excess unbound antibody was washed off by 5 rinses of 5 minutes each, in 10 mM glycine/PBS. A secondary horse anti-mouse FITC-conjugated antibody (Vector Lab., U.K.) was added in blocking solution at a final dilution of 1/1000 and left on the cells for 2-12 hours at room temperature. During this incubation, the cells were protected from light to prevent fading of the fluorescein reagent. After washing, the coverslips were mounted onto glass slides using an antifade mountant.

Immunofluorescence Analysis

Cells were analysed by phase contrast microscopy and by fluorescence with 495 nm UV light excitation. A Silicon Intensified Target (SIT) camera coupled to an Argus 100 image processor (Hamamatsu Photonics) were used for quantitative image analysis. The SIT camera gives real time imaging at television frame rates (40 milliseconds). The fluorescent images were derived from integrations of 64 frames.

RESULTS

TAT48 is Unable to Activate HIV-1 Directed Gene Expression in COS-7 Cells

To ask whether TAT48 could function in mammalian cells, we cotransfected COS-7 cells with a TAT48 expression vector and the HIV-1 LTR CAT reporter plasmid pOGS210. The TAT expression plasmid contained an SV40 origin of replication which allows the plasmid to replicate to a high copy number in COS-7 cells due to the expression of the SV40 large T antigen in these cells. This insures high levels of protein expression, allowing easier immunodetection of the expressed proteins. As a positive control an expression vector encoding the full-length TAT protein was also used. Forty-eight hours after transfection, protein extracts were assayed for CAT activity. No CAT activity was found from transfections with the TAT48 expression vector (Figure 2 K), whereas (Figure 2 J) a high level of activity was found from transfections with the full-length TAT expression vector.

TAT48 is Predominantly Found in the Cytoplasm of Transfected COS-7 Cells

To assess whether the lack of transactivation activity of the TAT48 protein might be due to the loss of nuclear/nucleolar localisation, we performed indirect immunofluorescence analysis on transfected COS-7 cells. We used as primary antibody a mouse monoclonal primary antibody raised against the N-terminal 15 amino acids of the TAT protein and as secondary antibody an FITC-conjugated horse antibody raised against mouse immuno-globulin-G. TAT48 was predominantly localised in the cytoplasm of COS-7 cells (Figure 2 B, E and H) whereas TAT was almost entirely in the nucleus (Figure 2 A, D and G). Image analysis performed on 10 individual transfected cells suggested that an average of 68% of TAT48 protein was in the cytoplasm and 97% of TAT in the nucleus (Figure 2 G and H).

C-MycTAT48 Protein is Unable to Activate Gene Expression from the HIV-1 LTR Despite Being Predominantly Localised in the Nucleus

One explanation for the lack of transactivation activity of the TAT48 protein could be due to its cytoplasmic localisation. In order to investigate the importance of nuclear localisation for transactivation, we constructed an expression vector which encoded the nuclear localisation sequence (NLS) of the oncogene C-Myc fused to TAT48. This fusion protein should be relocated to the nucleus due to the presence of the C-Myc NLS. We found that, as with the wild type TAT protein, a large proportion of the protein (75%) was now localised in the nucleus (Figure 2 C, F and I). However, unlike the TAT protein, the C-MycTAT48 protein was excluded from the nucleolus (6%, Figure 2 I). This agreed with the previous finding that the C-Myc NLS is able to target the cytoplasmic ß-galactosidase protein to the nucleus but not to the nucleolus (Dang *et al.,* 1988). The C-MycTAT48

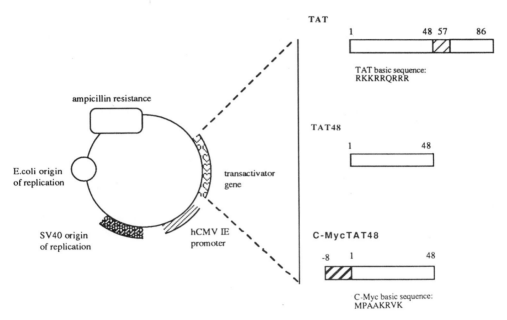

Figure 1. Diagram of the expression vector encoding the various TAT-related transactivators.

fusion protein, like TAT48, was not active in transactivation experiments (Figure 2 L). This indicates that the C-Myc NLS was not able to replace the function of the C-terminal region of TAT which is lacking in TAT48 (amino acids 49-86), even though the C-MycTAT48 protein is targeted to the nucleus (Figure 2 C, F and I). These observations could be explained by two hypotheses. Either a) the C-MycTAT48 protein is not able to interact specifically with TAR RNAs and that this specific TAT-TAR interaction is required for activation, or b) the loss of activity is due to the absence of the C-MycTAT 48 protein from the nucleolus, where 29% of the TAT protein is found. This would imply that the localisation of the activator protein to the nucleolus is required for the transactivation activity.

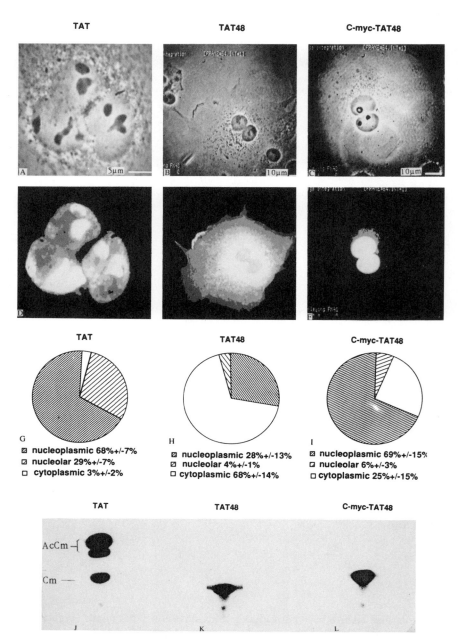

Figure 2. Quantitative analysis of the cellular localisation and transactivation activity of various TAT related proteins in COS-7 cells. **A-C.** Phase contrast images (bars show relative size). **D-F.** Fluorescence images. **G-I.** Schematic representations of the quantitative localisation of proteins in different cellular compartments. Results from pools of 10 cells as % total fluorescence ± standard deviation. **J-L.** CAT assays (AcCm = acetylated derivatives of ^{14}C chloramphenicol, Cm = unconverted ^{14}C-chloramphenicol).

Analysis of the Quantitative Localisation Results

The use of imaging analysis to quantify protein localisation was particularly useful for the experiments described above. However, for this type of quantitative analysis it is particularly important to be able to assess the specificity of the method and to identify any possible artefacts and how they might be avoided. As a control for our quantification data, we performed a similar analysis for two proteins of known subcellular localisation: SV40 large T antigen and ß-galactosidase. The SV40 large T antigen is constitutively expressed in COS-7 cells and has previously been shown to be specifically localised within the nucleus, but excluded from the nucleolus. The ß-galactosidase protein is a cytoplasmic protein, which was expressed after transfection with a ß-galactosidase encoding plasmid. Both proteins were detected using polyclonal rabbit primary antibodies and an FITC-conjugated goat anti-rabbit secondary antibody. About 80% of the SV40 large T antigen was found in the nucleus and only 7% in the nucleolus (data not shown). This agreed with the previous observed localization (Kalderon et al., 1984) and also with the results obtained from our C-MycTAT48 construct which contains a similar NLS. For the cytoplasmic ß-galactosidase, 40% of the protein was found in the nucleus and 2% in the nucleolus (data not shown). The percentage of fluorescence found in the nucleus is high for a known cytoplasmic protein, although this result is similar to that obtained for the TAT48 protein. There are two possible explanations for such an apparently high nuclear accumulation. Firstly, COS-7 cells have the ability to extensively replicate plasmids harbouring an SV40 origin of replication and therefore overproduce the encoded proteins. A cytoplasmic protein could therefore passively diffuse to the nucleus. Such a passive diffusion has previously been observed in *Xenopus* oocytes when increasing amounts of the cytoplasmic bovine serum albumin was injected into the cytoplasm and subsquently found in the nucleus (Goldfarb et al., 1986). However, although such a passive diffusion might be possible for the relatively small TAT48 protein (approximately 8 kD), it is not likely for the larger ß-galactosidase protein (approximately 116 kD) since proteins of this molecular weight have been shown to be unable to diffuse across the nuclear membrane (Paine et al., 1975). Another possibility is that the observed nuclear fluorescence results from cytoplasm overlaying the nucleus. We also observed that there is a certain degree of variability in our localisation results (Figure 2 G, H and I). We currently quantified 10 transfected cells for each protein. Although this is insufficient for proper statistical analysis, we observed that the degree of variability is low for the nuclear TAT protein (97% $\pm$ 2% nuclear localisation) but higher for the cytoplasmic TAT48 (30% $\pm$ 15% nuclear localisation). This difference could indicate two distinct mechanisms for the nuclear accumulation of the two proteins: a passive diffusion for the cytoplasmic TAT48 and a controlled, active transport for the full-length TAT protein. Passive diffusion might result in different subcellular distributions depending on the amount of expressed protein, whereas an active transport should lead to a constant subcellular distribution regardless of the amount of synthesised protein.

Expression of Lower Levels of TAT Proteins in HeLa Cells

The COS cell system is commonly used for the over-expression of proteins. The data discussed above raised the possibility that this over-expression of the TAT proteins might give rise to artefacts in the localisation results. In HIV-1 infected cells, the levels of expression of the transactivator proteins is very low (Haseltine, 1991). Therefore, it was important to study the localisation of the activator proteins at the lowest levels of expression which were reasonably detectable. Therefore, we used HeLa cells in which the expression

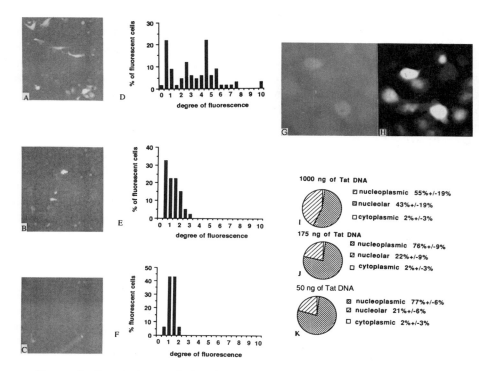

Figure 3. Quantitative analysis of TAT localisation in HeLa cells following transfection with different concentrations of DNA. **A-C**. Immunofluorescence localisation of HeLa cells expressing TAT protein following transfection with different DNA concentrations. (A. 1000 ng, B. 500 ng, C. 175 ng). **D-F**. Quantitative analysis of the data shown in images A-C. showing the relative numbers of fluorescent cells at different intensities. **G** and **H**. Comparison of detection of TAT protein in HeLa cells with a standard 35 mm camera and with the electronic SIT camera. The HeLa cells were transfected with 50 ng DNA. **I-K**. Schematic representation of the quantitative results of intracellular localisation of TAT protein following transfection of HeLa cells with different concentrations of DNA.

plasmids do not replicate. In order to further reduce the level of the protein expression per cell we titrated down the concentration of the transfected DNA. Interestingly, when large amounts of DNA (1000 ng/10^4 cells) were transfected, a large variation in the intensity of fluorescence of the transfected cells was observed (Figure 3 A and D). The degree or intensity of fluorescence was defined as the total fluorescence counts per unit of area. In the latter case, we observed a highly heterogenous population which ranged from 0.5 to 10.5 standard units. This is a measure of the efficiency with which the cells take up DNA and express protein. When lower amounts of transfecting DNA were used (500 ng and 175 ng/10^4 cells) this variability was considerably reduced (Figure 3 B, C, E and F).

Analysis of the Subcellular Localisation of TAT in HeLa Cells Transfected with Different DNA Concentrations

We investigated whether the TAT subcellular localisation was affected by the concentration of transfected DNA. As was previously found in COS cells, 98% of TAT was found in the nucleus with a variation of 3% (Figure 3 I, J and K). This was found to be independent of the amount of input DNA. Interestingly, the percentage of TAT found in the nucleolus was in some way related to the amount of transfecting DNA. At high DNA concentrations (1000 ng/10^4 cells) 45% of the TAT protein found in the nucleus was present in the nucleolus, whereas at lower DNA concentration (175 ng/10^4 cells) only 20% of the nuclear TAT protein was found in the nucleolus. This percentage was the same at the concentration of 50 ng/10^4 cells. This shows that a significant percentage of TAT (20%) is found in the nucleolus even when low amounts of protein are present.

DISCUSSION

We have shown that the wild-type 86 amino acid length TAT protein is specifically localised in the nucleus of transfected COS and HeLa cells. The protein also appears to be concentrated within the nucleoli. These results agree with those described previously (Hauber *et al.*, 1987; Ruben *et al.*, 1989; Siomi *et al.*, 1990). A 48 amino acid length protein, TAT48, derived from deletion of the 38 C-terminal amino acids of TAT had no activity in transactivation experiments in mammalian cells and was no longer concentrated in the nucleus. This protein appeared to be excluded from the nucleolus. This deletion removes a small cluster of 9 basic residues from amino acids 49 to 57 (RKKRRQRRR) which constitutes the previously defined (Siomi *et al.*, 1990) TAT nuclear/nucleolar localisation signal (see Figure 1). A basic sequence extended by 3 C-terminal amino acids (RRKKRRQRRRHQN) has been implicated in defining nucleolar localisation (Siomi et al., 1990). TAT proteins, which have deletions of the C-terminus that remove 2 of the 3 amino acids of this nucleolar localisation sequence, have been shown to have significant but reduced transactivator activity (Seigel *et al.* 1986). In our previous results, the TAT48 protein was fully active when injected into the nucleus of *Xenopus* oocytes (Braddock *et al.*, 1989). This implies that the TAT48 protein retains at least part of the transactivation domain of TAT. One possible explanation of the present results was that the removal of the basic sequence gave rise to inappropriate cytoplasmic targeting of the TAT48 protein in mammalian cells, so that the protein was unable to activate gene expression. Therefore, we constructed the C-MycTAT48 fusion protein expression vector in which the C-Myc nuclear localisation sequence was fused at the N-terminus of the TAT48 sequence. Expression of this protein in mammalian cells gave nuclear but not nucleolar localisation. The C-MycTAT48 protein had no transactivator activity in mammalian cells. This result indicates that simple nuclear localisation of the C-MycTAT48 protein is not sufficient to give TAT activity in mammalian cells. The TAT basic sequence has also been shown to be essential for the binding of the TAT protein to TAR RNA *in vitro* (Dingwall *et al.*, 1989; Dingwall *et al.*, 1990; Kamine *et al.*, 1991; Weeks *et al.*, 1990). How then can TAT48 be functional in *Xenopus* oocytes when the previously characterised TAT binding site for TAR has been deleted? These results clearly imply a complex role for different overlapping domains of the TAT protein which may have key roles in the different and interrelated processes of cellular localisation, RNA binding and activation of transcriptional and translational gene expression. A key question from these results is whether localisation to the nucleolus is essential for TAT action. The C-MycTAT48 protein was not concentrated into the nucleoli in mammalian cells as is the case with full length TAT protein. The apparent high level of TAT accumulation into the nucleoli of mammalian cells has been shown above to occur both at high levels of TAT expression, but also when the level of TAT expression is considerably reduced. This may suggest a functional significance for the nucleolar localisation of TAT. It is important for accurate immunofluorescent quantification to investigate the various roles of cell fixation procedure, choice of fluorophore and conditions for the illumination and microscopic examination of the sample. We have reproduced the above results with antibodies labelled with rhodamine (data not shown). A further important factor described above, is the requirement to ensure that the protein of interest is expressed within the cells at a physiologically relevent concentration. Examination and controls for each of these factors will ensure against artifactual results due to experimental conditions which have no significance in the living cell. An important tool in the studies described above has been the use of digital imaging technology. This has permitted the quantification of the intracellular localisation of proteins by the use of immunofluorescence. As well as the potential for quantification, we have shown that the use of the SIT camera permitted the detection of fluorescence with greater sensitivity and improved image contrast (see Figure 3). We have also found that the use of the SIT camera, which has an intrinsic horizontal resolution of 500 TV lines, gives rise to pictures with an apparent intracellular resolution which is very close to that obtained with a 35 mm camera and photographic film (data not shown). Due to the wide wavelength sensitivity of these cameras (~ 400 nm to 850 nm), they are useful for fluorescent imaging with a wide range of different fluorophores. In addition, this camera has a dynamic range covering three orders of magnitude (10^{-1} to 10^{-4} lux). The ability of digital imaging systems to quantitatively localise ion concentrations, reporter enzymes (White *et al.*, 1990) proteins and nucleic acids within cells has many implications for future biological studies at the single cell level.

ACKNOWLEDGEMENTS

We would like to thank Hamamatsu Photonics for the loan of the SIT camera and Argus 100 image processor, and British Biotechnology for financial support.

REFERENCES

Adams, S., Johnson, J., Braddock, M., Kingsman, A.K., Kingsman, S.M., and Edwards, M. (1988) Synthesis of a gene for the HIV transactivator Tat by a novel single stranded approach involving *in vivo* gap repair. *Nucl. Acid Res.* **16**: 4287-4298

Braddock, M., Chambers, A., Wilson, W., Esnouf, M.P., Adams, S.E., Kingsman, A.J., and Kingsman, S.M. (1989) HIV-1 TAT activates presynthesised RNA in the nucleus. *Cell* **58**: 269-279

Braddock, M., Thorburn, A.M., Chambers, A., Elliott, G.D., Anderson, G.J., Kingsman, A.J., and Kingsman, S.M. (1990) A nuclear translation block imposed by the HIV-1 U3 region is relieved by the TAT-TAR interaction. *Cell* **62**: 1123-1133

Braddock, M., Thorburn, A.M., Kingsman, A.J., and Kingsman, S.M. (1991) Blocking of TAT dependent HIV-1 RNA modification by an inhibitor of RNA polymerase II processivity. *Nature* **350**: 439-441

Dang, C.V., and Lee, W.M.F. (1988) Identification of the human c-myc protein nuclear translocation signal. *Mol. Cell. Biol.* **8**: 4048-4054

Dayton, A.I., Sodroski, J.G., Rosen, C.A., Goh, W.C., and Haseltine, W.A. (1986) The transactivator gene of the human T-cell lymphotropic virus type III is required for replication. *Cell* **44**: 941-947

Dingwall, C., Ernberg, I., Gait, M.J., Green, S.M., Heaphy, S., Karn, J., Lowe, A.D., Singh, M., Skinner, M.A., and Valerio, R. (1989) Human immunodeficiency virus 1 tat protein binds trans-activation-reponsive region (TAR) RNA *in vitro*. *Proc. Natl. Acad. Sci. USA* **86**: 6925-6929

Dingwall, C., Ernberg, J., Gait, M., Green, S., Heaphy, S., Karn, J., Lowe, A., Singh, M., and Skinner, M. (1990) HIV-1 TAT protein stimulates transcription by binding to a U-rich bulge in the stem of the TAR RNA structure. *EMBO J.* **9**: 4145-4153

Fisher, A.G., Feinberg, M.B., Josephs, S.F., Harper, M.E., Mansell, C.M., Reyes, G., Gonda, M.A., Aldovini, A., Debouk, C., Gallo, R.C., and Wong-Staal, F. (1986) The transactivator gene of HTLV-III is essential for virus replication. *Nature* **320**: 367-371

Goldfarb, D., Gariepy, J., Schoolnik, G., and Kornberg, R.(1986) Synthetic peptides as nuclear localisation signals. *Nature* **322**: 641-644

Gorman, C. (1982) High efficiency gene transfer into mammalian cells, *in:* "DNA Cloning, Volume II: A Practical Approach", D.M. Glover, ed., IRL Press, Oxford,143-165

Haseltine, W.A. (1991) Regulation of HIV-1 replication by tat and rev, *in:* "Genetic Structure and Regulation of HIV", W.A Haseltine and F. Wong-Staal, eds., Raven Press Ltd., New York, 1-42

Hauber, J., Malim, M., and Cullen, B. (1987) Mutational analysis of the conserved basic domain of human immunodeficiency virus tat protein. *J. Virol.* **63**: 1181-1187

Jaeng, K.-T., Shank, P.R., and Kumar, A. (1988) Transcriptional activation of homologous viral long terminal repeats by the human immunodeficiency virus type I or the human T-cell leukaemia virus type I TAT proteins occurs in the absence of de novo protein synthesis. *Proc. Natl. Acad. Sci. USA* **85**: 8291-8295

Jakobovits, A., Smith, D.H., Jakobovits, E.B., and Capon, D.J. (1988) A discrete element 3 of the human immunodeficiency virus-1 (HIV-1) and HIV-2 mRNA initiation sites mediates transcriptional activation by an HIV transactivator. *Mol. Cell. Biol.* **8**: 2555-2561

Kalderon, D., Richardson, W.D., Markham, A.F., and Smith, A.E. (1984) Sequence requirements for nuclear location of simian virus 40 large-T antigen. *Nature* **311**: 33-38

Kamine, J., Lowenstein, P., and Green, M. (1991) Mapping of HIV-1 Tat protein sequences required for binding to TAR RNA. *Virology* **182**: 570-577

Maniatis, T., Fritsch, E.F., and Sambrook, J. (1990) "Molecular Cloning: A Laboratory Manual", Cold Spring Harbor Laboratory, New York

Muesing, M.A., Smith, D.H., and Capon, D.J. (1987) Regulation of mRNA accumulation by a human immunodeficiency virus transactivator protein. *Cell* **48**: 691-707

Paine, P.L., Moore, L. C., and Horowitz, S.B. (1975) Nuclear envelope permeability. *Nature* **254**: 109-114

Rice, A.P., and Mathews, M.B. (1988) Transcriptional but not translational regulation of HIV-1 by the tat gene product. *Nature* **332**: 551-553

Rosen, C.A., Sodroski, J.G., Goh, W.C., Dayton, A.I., Loppke, J., and Haseltine, W.A. (1986) Post-transcriptional regulation accounts for the trans-activation of the human T-lymphotropic virus type III long terminal repeat. *Nature* **319**: 555-559

Rosen, C.A., Terwilliger, E., Dayton, A., Sodroski, J.G., and Haseltine, W.A. (1988) Intragenic cis-acting art gene-responsive sequences of the human immunodeficiency virus. *Proc. Natl. Acad. Sci. USA.* **85**: 2071-2075

Roy, S., Delling, V., Chen, C.-H., Rosen, C.A., and Sonnenberg, N. (1990) A bulge structure in HIV-1 TAR RNA is required for TAT binding and TAT-mediated transactivation. *Genes Dev.* **4**: 1365-1373

Ruben, S., Perkins, A., Purcell, R., Joung, K., Sia, R., Bunghoff, R., Haseltine, W. A., and Rosen, C.A. (1989) Structural and functional characterization of human immunodeficiency virus tat protein. *J. Virol.* **63**: 1-8

Seigel, L.J., Ratner, L., Josephs, S.F., Derse, D., Feinberg, M.B., Reyes, G.R., OBrien, S.J., and Wong-Staal, F. (1986) Transactivation induced by human T-lymphotropic virus Type III (HTLV III) maps to a viral sequence encoding 58 amino acids and lacks tissue specificity. *Virology* **148**: 226-231

Siomi, H., Shida, M., Maki, M., and Hatanaka, M. (1990) Effects of a highly basic region of HIV Tat protein on nucleolar localisation. *J. Virol.* **64**: 1803-1807

Weeks, K., Ampe, C., Schultz, S., Steitz, T., and Crothers, D. (1990) Fragments of the HIV-1 tat protein specifically bind TAR RNA. *Science* **249**: 1281-1285

White, M., Braddock, M., Byles, E., Kingsman, A.J., and Kingsman, S.M. (1992) Application of the firefly luciferase reporter gene to microinjection experiments in *Xenopus* oocytes. This volume

White, M.R.H., Morse, J., Boniszewski, Z.A.M., Mundy, C.R., Brady, M.A.W., and Chiswell, D.J. (1990) Imaging of firefly luciferase expression in single mammalian cells using high sensitivity charge-coupled device cameras. *Technique* **2**: 194-201

CONCEPTS OF THE CYTOMETRIC APPROACH

Spencer C. Brown

Service de Cytométrie, Institut des Sciences Végétales, Centre National de la Recherche Scientifique, 91198 Gif-sur-Yvette, France

INTRODUCTION

There has been intense development of flow cytometric methods and techniques over the past 15 years, which constitute simple and routine procedures in many disciplines. Indeed, this chapter concludes with a brief guide to the literature, with a bias towards plant biology. But beyond such detail, the question remained. What is the conceptual significance, in terms of experimental design, of a technique which obtains a set of optical parameters from individual objects as they rapidly flow through an optical bench - with neither image nor twice the same object?

The answers to this question are a guide to the utility of flow cytometry in experimentation. Certainly the aspects which we shall discuss are not proprietary to flow cytometry. They do, however, underline desirable developments in sister techniques such as imaging. For instance, anticipating the first point below, namely the advantages of multiparametric analysis in cell biology, one may forecast a demand for three and four parameter confocal microscopy: this will require progress in (achromatic) optics, in fluorescent probes, possibly Fourier spectral deconvolution, and the computer techniques to represent multiparameter data on images for the scientist.

CONCEPTUAL FEATURES

Multiparametric Analyses with Logical Windows

The multiparametric nature of analyses linked by a chain of logical windows, gates or zones of interest, enables one to address a key observation to cells or other objects

Biotechnology Applications of Microinjection, Microscopic Imaging, and Fluorescence, Edited by P.H. Bach *et al.*, Plenum Press, New York, 1993

defined through other parameters. For instance, imagine cells labelled with two antibodies and for nuclear DNA content to position each cell within the cell cycle: multivariate association of these three characters is far more powerful than their three separate assessments. An example (June and Rabinovitch, 1988) is fluorometric analysis of cytoplasmic calcium flux in four B lymphocyte classes defined simultaneously with two surface antigens. A further example comes from Carayon et al. (1991). Using an antibody cocktail and immunofluorescence, they could simultaneously identify and quantify 8 leucocyte subsets of human peripheral blood. This single-laser (488 nm) analysis used three fluorochromes: fluorescein, phycoerythrin, and one of several red-emitting energy-coupled dyes with long Stokes' shift.

A third example of this strategy, which is fundamental to cytometry, comes from plant biology. Schröder and Petit (1992) used light scattering properties to classify isolated chloroplasts and thylakoids into subclasses (intact or not), and quantified their lectin binding properties with fluorescent ligands: the lectin result thus corresponded to "analytically purified" objects.

Large Sampling of Small Sample Size

Rare event analysis is possible with a technique measuring 3,000 objects per second. However, reliable statistics do not necessarily follow. It is insufficient to simply count many events to quantify residual metastatic bone marrow cells (frequencies 10^{-4}) or putative changes in polyploid cells (about 10^{-2}) in infected plants, because false positives or noise must be diagnosed. In general, one has to resort to a multiparametric approach, finding secondary markers to eliminate these false positives from the key parameter.

We have sorted Madagascan periwinkle protoplasts (*Catharanthus roseus*) at a level of 10^{-3}, based on their exceptional endogenous blue fluorescence, to establish cultures from rare cell types with respect to indole alkaloid metabolism (Brown et al., 1984; Bariaud-Fontanel et al., 1988).

Single cell biochemistry with fluorescent probes also offers an economy of biological material. This is demonstrated in a screen of 16 cell functions (Valet and Roth, 1991), under clinical trials in Germany for early diagnosis and prognosis of septicaemia. Only 18 years ago a DNA assay had a sensitivity to 10^{5} nuclei, whereas current flow cytogenetics have a resolution approaching 10^{-3} of the genome (Gray and van der Ploeg, 1990).

Process Raw Samples

Working with unpurified samples can be simpler, cleaner (e.g. reduced manipulation of blood samples), and may retain the biological heterogeneity which is otherwise altered during preparative procedures such as centrifugation. Galbraith et al. (1983) introduced a simple chopping method to isolate nuclei for total DNA analysis. Figure 1 and colour plate 3 provides such an example. After several minutes for staining, filtration and cytometry, a highly resolved histogram of total nuclear DNA is obtained. In difficult tissue such as roots, a typical histogram of 10,000 nuclei will still retain its precision, with a coefficient of variation below 3% on each peak, although several hundred thousand objects other than nuclei (starch, plastids, membrane and wall fragments) have gone through the analysis. Where purification from plastidial contamination and cell-wall alginates would have been difficult, algal genomes have been quantified through simple cytometric methods (Le Gall *et al.*, 1992).

There are several industrial applications of DNA cytofluorometry (Brown *et al.*, 1991b; Brown, 1992). Micro-organism counts and viability are rapidly derived in yoghurt and other milk products through flow cytometry (Dumain *et al.*, 1990).

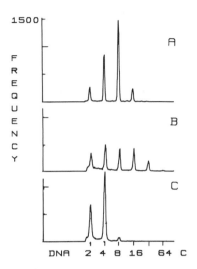

Figure 1. Monoparametric histograms of DNA content in 10,000 nuclei isolated from (**A**) rosette leaf, (**B**) its petiole, or (**C**) roots of *Arabidopsis thaliana*, showing the typical level of polysomaty. The tissue was simply chopped with a razor blade, stained with bisbenzimide Hoechst 33342, filtered through a 30 μm mesh, and analysed on an EPICS 751 flow cytometer (Coultronics, Margency, France). See Brown *et al.* (1991a) for methods. The fluorescence emission was amplified on a logarithmic scale, and calibrated to C-values (*A. thaliana* 2C = 0.34 pg) with an internal standard (chicken erythrocytes, 2C = 2.33 pg DNA, not shown here). The ordinate scale is constant. The endoreplications are developmentally regulated, patterns being identical for the same tissue on different plants. The various ploidy populations have essentially the same variation, with coefficients of variation about 3%. These histograms were conditioned upon other parameters i.e. the 10,000 root nuclei were analysed along with 300,000 other irrelevant objects (debris, doublets, starch, etc.), these being excluded by logical windows.

Communicating Multiparametrically

One of the challenges for many modern analytical techniques is how to communicate multidimensional data to the biologist. A histogram is one parameter against frequency. The term "cytogram" has come into use for a bi-variate histogram, commonly displayed as grey-levels or frequency contours on coordinate axes (e.g. Figures 2 and 4), occasionally as a pseudo-three dimension representation with Z as frequency. Three cell parameters may be represented on triple axes, with frequency appearing as clusters or clouds. Beyond this, mathematical compression may be performed, such as reduction of parameters to ratios. One thing is clear, colour is a powerful aid to navigation through multivariate data, where populations determined on one set of axes are assigned colours which identify the same objects on the representations of the other parameters. Terstappen *et al.* (1990) have provided an elegant example of this routine practice.

Returning to confocal microscopy, consider a computer image of a cell with dual immunofluorescence labelling. There is a need to represent the statistics of the interaction or colocalisation of the dual labelling (and not just the respective grey-levels) upon this image. Davoust (from Theuillon *et al.*, 1991) has addressed this problem by representing the dual pixel values on X,Y coordinate axes as a bi-variate histogram, with statistics of co-localisation. Zones of interest within this statistical map can then be assigned false colours, promptly displayed at the relevant pixels in the original cell image. Pixels with high fluorescein may be represented in green, those with Texas Red in red, pixels with high intensities of both are given blue, those with low intensities of both are yellow etc. The choices are arbitrary and interactive, far more incisive than the simple complementation of colour intensities.

Physical Separation

The possibility of physically sorting the analysed object, developed with flow cytometry in the late 1970s, remains a useful option. Cells, chromosomes etc. may be collected for culture, biochemistry or microscopy. The Los Alamos National Laboratory has sorted from flow karyotypes for more than 1000 chromosome banks. Several microscope-based laser sorting techniques are also available, particularly relevant now that even smaller quantities of biological material are needed for culture or for processing with PCR. Moving particles with a laser beam as an optical trap is an exciting development in this field (Greulich, 1992).

Plant nuclei stained for DNA may be sorted on a preparative scale by flow cytometry as a simple and elegant way of obtaining "cycle synchronized" nuclear material (Perennes *et al.*, 1990). The sorted product can then be analysed for specific mRNA (Bergounioux *et al.*, 1992) or for viral DNA (Accotto *et al.*, 1991) at about 1000 nuclei per second.

Intrinsic Fluorescence

Intrinsic fluorescence might serve as an indicator of cell physiology or classification, and in any case will be a major consideration when planning fluorometric experiments as shown in colour plates 4 and 5.

Plant tissues generally contain several strong fluorophores (Jameson and Reinhart, 1989). Chlorophylls, with their red emission, are a classic example but not necessarily the most dominant. Indeed, most tissues contain blue emitting compounds: nicotinamide nucleotides (NADPH emission max 450 nm), alkaloids (e.g. serpentine emission max 430 nm), hydroxycinnamic acid derivatives such as *p*-coumarins and ferulic acid esterified with polysaccharides of the primary cell walls, and pteridines (emission 440 nm). Flavins will emit around 535 nm.

Much as *in situ* chlorophyll fluorescence has served as an indicator of photosynthetic status, other autofluorescence might provide intrinsic probes of cell physiology. Goulas *et al.* (1990) performed time-resolved deconvolution of plant autofluorescence in a program to use remote sensing strategies for stress detection in agriculture. Other reports have addressed pollution and phytotoxin effects (Berglund *et al.*, 1988; Sigal *et al.*, 1988). An example concerning water stress in lupin is detailed below

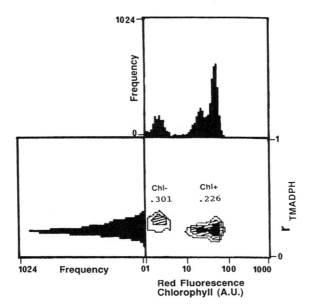

Figure 2. Cytogram corresponding to 6,000 protoplasts of a drought-resistant lupin genotype, each analysed for red chlorophyll (Chl, log scale) and TMA-DPH emission anisotropy (r, linear). Three classes of frequency are indicated by the contours which delimit coordinates having ≥ 35 protoplasts (solid black), 12-34 protoplasts (hatched), 4-11 protoplasts (white). Corresponding histograms are shown as projections. The monomodal anisotropy histogram in fact confounds distinctive anisotropy values for Chl⁻ and Chl⁺ protoplasts; the rTMA-DPH values shown are the means of each class as delimited in the bivariate analysis. The two contiguous Chl⁻ subclasses have similar modal anisotropy, and results elsewhere indicate that mesophyll protoplasts with only traces of chlorophyll still have this same anisotropy (from Gantet *et al.*, 1990, with permission).

(Gantet *et al.*, 1990). Oceanographers use pigment fluorescence as signatures for species or classes of organisms in flow cytometry of sea water (Yentsch and Horan, 1989).

With natural blue fluorescence as a marker for protoplast integrity, Matine (1991) sorted out the rare chlorophyll-free protoplasts (nul red fluorescence) from suspensions derived from Citrus epidermis and, with their lysis, prepared pure leucoplasts. C10-monoterpene synthesis was then assigned to these leucoplasts on the basis of radioisotope studies with the purified fractions.

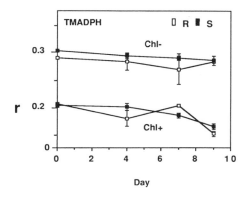

Figure 3. Emission anisotropy of mesophyll (Chl$^+$) or chlorophyll-less (Chl$^-$) subpopulations after TMA-DPH labelling of lupin protoplasts of the resistant (R, open symbols) or susceptible (S, solid) genotypes. The plants had undergone 0, 4, 7, or 9 days without watering. The 95% confidence intervals do not exceed the symbols except where shown (Gantet *et al.*, 1990).

What to Do When Intrinsic Fluorescence is an Unwanted Background?

Firstly, it should be characterized by spectrofluorometry, under the various conditions of the experiment, such as with and without stress. Pre-treatments such as photobleaching or etiolation may be envisaged.

The extrinsic fluorescent marker (e.g. the fluorescent label for an antibody) will then be selected with a view to minimizing interference at excitation and emission wavelengths.

Narrow band filters are usually justified because the priority is normally to favour specificity of a signal reaching a photodetector rather than its intensity. Unfortunately, too many epifluorescence microscopes are purchased without due investment in specific bandpass observation filters.

Pulse compensation techniques will improve specificity. This refers to the practice of measuring two fluorochromes at their respective optima, and reciprocally correcting the cell signal of each channel for the minor contribution of the other within its spectral window. For example, when quantifying a fluorescein labelled antibody on mesophyll protoplasts, the fluorescein emission (I_f) may be read at 520 ± 10 nm, chlorophyll (I_c) at 685 ± 5 nm, and for each cell the corrected intensities (I') are calculated:

$$I'_f = I_f - a.I_c \qquad\qquad (1)$$

$$I'_c = I_c - b.I_f \qquad\qquad (2)$$

where the coefficients a and b have been measured as the fractional overlap of the two spectra in this optical configuration. Single-labelled cells are run initially to assess this level of spectral cross-talk. Pulse compensation is a routine function of cytometry with two or

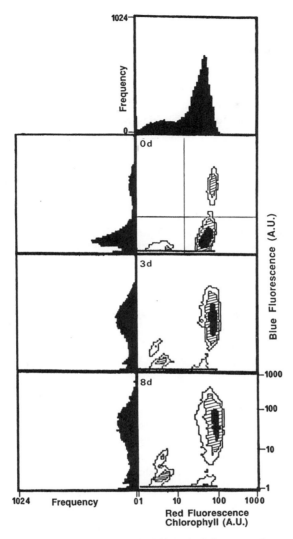

Figure 4. Cytograms corresponding to 20,000 lupin foliar protoplasts each analysed for red chlorophyll and natural blue fluorescence. The axes (64 x 64 channel resolution) are logarithmic. Three classes of frequencies are indicated by the contours which delimit coordinates having ≥ 30 protoplasts (solid black), 10-29 protoplasts (hatched), 5-9 protoplasts (white). Statistics for each class were obtained with a typical quadrant analysis (X,Y : 25,20) demonstrated on the first cytogram. Corresponding histograms are shown as projections for the blue fluorescence, and for the red fluorescence only on day 0. The protoplasts were from plants of the resistant genotype, after day 0, 3 or 8 without watering. Stress recruited cells to the class of strong blue fluorescing protoplasts (the highest frequency contour that is reached) but had little incidence upon their modal fluorescence intensity as its Y coordinate was stable (Gantet *et al.*, 1990).

three colour immunofluorescence. It is one way of correcting for autofluorescence, providing this has some distinctive spectral features either at the excitation or emission wavelengths. Multidetector confocal and imaging microscopes offer similar pulse compensation functions applied pixel by pixel (rather than object by object) to reduce spectral cross-talk.

The spatial resolution of imaging techniques will often allow a better signal to noise ratio than in cytometry. A typical example is assessment of *in situ* hybridisation on nuclei. Measurement of a hybridisation spot is compromised in flow cytometry by the total background "noise" of the nucleus, whereas imaging techniques, localising the spot, have a simpler signal/noise problem. Conventional quantitative fluorescence microscopy has been difficult on plant material. Fortunately, confocal microscopy appears to be filling this lacuna, with even modest spatial resolution it is a boon to quantitative plant microscopy in intact or living cells.

Time resolution is another strategy to lessen the contribution of autofluorescence which often has a short excited-state half-life. Time-resolved microscopes and flow cytometers are being developed in association with long-lifetime probes incorporating rare-earth ions.

Vital Probes

Understanding the dynamics of biological processes and multivariate relationships has been facilitated by the development of non-invasive techniques: for this subject, Forskett and Grinstein (1990) reads like a novel. Other chapters of the present book provide examples (see those by Bach *et al.*; Hermetter *et al*; Horobin and Rashid-Doubell; Rashid-Doubell and Horobin). Vital fluorescent probes form part of this arsenal, but much innovation is still required. Non-destructive methodologies should be developed for vital DNA stains to classify plant protoplasts or cells with respect to the cell cycle (bisbenzimide Hoechst 33342 is valid for mammalian cells), for reporter gene activity (e.g. lacZ, GUS), enzyme activities, intracellular antigens, transduction signals, longterm cell tracking (Horan and Slezak, 1989), to name just a few. The two companies that have substantially contributed to the development and availability of fluorescent and vital probes are Molecular Probes (Pitchford Avenue, Eugene, Oregon, USA) and Lambda Fluoreszenztechnologie (Gottenhof-Strasse, Graz, Austria).

Fluorimetry associated with sorting offers the possibility of not only studying the cell or organelle, but of retaining it for culture.

Molecular Specificity and Quantification

Cytologists developed tools offering improved molecular specificity to meet the potential of, and demands on, this technique. Initially one used protein-reactive dyes and then immunofluorescent techniques were developed. Initially one assessed genome size by DNA length, then base-pair composition with specific dyes (for calculation procedures, see Godelle *et al.*, 1992), then multi-fluorescence *in situ* hybridisation, and now real-time analysis of nucleic acid processing (T. Montenay-Garestier, Institut Curie, Paris: see Mergny *et al.*, 1992).

Fluorometry generally uses arbitrary units. Greater use of internal references and innovation in calibration methods is warranted.

Time

Imaging techniques have, at last, facilitated the cell biologists' study of the dynamics of cell processes. In contrast, flow cytometers capable of providing kinetic information are unfortunately rare.

AN EXAMPLE

A cytometric approach to handle biological heterogeneity (Figures 2-4 and plates 4 and 5) is shown by comparing the cellular properties of drought resistant or susceptible lupin genotypes (Gantet *et al.* 1990) by labelling the plasma membrane of leaf protoplasts with lipophilic 1-(4-trimethylammonium)-6-phenyl-1,3,5-hexatriene p-toluene sulfonate (TMA-DPH), and assessing steady-state emission anisotropy, **r**. This parameter is related in a nonlinear manner to the dynamic property microviscosity and to structural order of the lipid matrix. The term "lipid fluidity" is used to englobe the two aspects.

Analogous biophysical studies have been undertaken with purified membrane preparations under tightly controlled temperature and with time-resolved fluorescence (see review Aloia *et al.* 1988). This was certainly not the situation for Gantet's cytometric study using whole protoplasts. However, the lupin protoplast suspension included at least two cell types, epidermal and mesophyll, which could be identified by their intensity of red chlorophyll fluorescence. Their plasma membrane viscosities were vastly different (Figure 2). As the suspension flowed object by object through the cytometer, light-scatter, red and blue fluorescence, and the anisotropy of this blue emission were calculated. The cytometric values of anisotropy therefore corresponded to the plasma membrane specifically labelled in intact protoplasts, mesophyll on the one hand and epidermal on the other. Broken cells had internal membranes labelled; they were identified by light-scattering properties and excluded from the final result by logical windows. Similarly, debris particles such as plastids were excluded. Any bulk fluorometric method, in a cuvette, cannot offer such discrimination. Furthermore, any treatment changing the frequency of the different protoplast classes (or of the debris) will change the global mean although cell parameters may be stable. Any membrane purification procedure will suffer from uncertainty not only with respect to membrane purity or stability, but regarding the cell source. The variance in Figure 3 is small, as shown by the tight error bars, but the major factor of membrane structure difference, that of cell type, was hitherto ignored. Epidermal cells have higher lipid microviscosity and/or more ordered lipid structure. Water stress greatly increased the apparent fluidity, and more so in the susceptible genotype; the effect was more pronounced in the chlorophyll-containing mesophyll cells than in the epidermal cells. Moreover, lupins accumulate alkaloids under stress. Figure 4 shows that some cells (5.4% of protoplasts at day 0) were rich in natural blue fluorescing compounds (about 15-fold more intense) even before the drought treatment. Stress increased the frequency of strong blue fluorescing protoplasts, having no incidence upon the modal fluorescence intensity of this class.

FURTHER READING

Melamed *et al.* (1990) provide the most substantial current presentation of flow cytometry. Crissman and Darzynkiewicz (1989) provide many methodologies.

Plant flow cytometry has also been reviewed by Brown and Bergounioux (1989) and Galbraith (1989). Methodologies for plant DNA analysis have been collated by Brown *et al.* (1991a) and Dolezel (1991), to which should be added the high-fidelity DNA analysis of Ulrich and Ulrich (1991) and the 2C DNA values for 130 species (Arumuganathan and Earle, 1991). Industrial and research uses of nuclear DNA assessment, notably for routine and massive ploidy determination, are discussed in Brown *et al.* (1991b) and Brown (1992), the latter imagining "bush DNA cytometry" such that the cytometer might go to the plant rather than the samples coming to the laboratory. Plant flow karyotyping is briefly reviewed in Veuskens *et al.* (1992), a report on sorting of Y-chromsomes from the dioecious plant *Melandrium album*. Protoplast sorting is treated by Harkins and Galbraith (1987), Lister (1989) and Hammatt *et al.* (1990), pollen microspores by Deslaurier *et al.* (1991). Bergounioux *et al.* (1982) addresses protoplast regeneration and cell proliferation. Studies on isolated mitochondria and chloroplasts are treated in Petit *et al.* (1989), Xu *et al.* (1990) and Petit (1992).

ACKNOWLEDGEMENTS

The author has appreciated the support of the European Economic Community (BRIDGE contract BIOTEC 0206F) to attend the BAMMIF Workshop.

REFERENCES

Accotto, G.P., Mullineaux, P.M., Brown, S.C., and Masenga, V. (1991) Digitaria streak geminivirus replicates in S-phase cells of *Digitaria setigera, in:* "International Symposium Plant Molecular Biology", R.B. Hallick, Ed., Tucson, Arizona, Abstract 1136

Aloia, R.C., Curtain, C.C., and Gordon, L.M. (1988) "Advances in Membrane Fluidity", AR Liss, New York

Arumuganathan, K., and Earle, E.D. (1991) Nuclear DNA content of some important plant species. *Plant Mol. Biol. Rep.* **9**: 208-218

Bariaud-Fontanel, A., Jullien, M., Coutos-Thevenot, P., Brown, S.C., Courtois, D., and Petiard, V. (1988) Cloning and cell sorter, *in:* "Plant Cell Biotechnology," M.S.S. Pais, ed., Springer Verlag, Heidelberg, 403-419

Berglund, D.L., Strobel, S., Sugawara, F., and Strobel, G.A. (1988) Flow cytometry as a method for assaying the biological activity of phytotoxins. *Plant Science* **56**: 183-188

Bergounioux., C., Brown, S.C., and Petit, P. (1992) Flow cytometry and plant protoplast cell biology. *Physiol. Plant.* **85**: 374-386

Brown, S.C. (1992) Cytométrie tout terrain or bush DNA cytometry, *in:* "New Developments in Flow Cytometry", A. Jacquemin-Sablon, ed., NATO ASI Series Volume, Springer Verlag, Heidelberg (in press)

Brown, S.C., and Bergounioux, C. (1989) Plant flow cytometry, *in:* "Flow Cytometry: Advanced Research and Clinical Applications." A. Yen, ed., vol II. CRC Press, Boca Raton Florida, 195-220

Brown, S.C, Bergounioux, C., Tallet, S., and Marie, D. (1991a) Flow cytometry of nuclei for ploidy and cell cycle analysis, *in:* "A Laboratory Guide for Cellular and Molecular Plant Biology," I. Negrutiu, and G.B. Gharti-Chherti, eds., Birkhäuser Verlag, Basel, 326-345

Brown, S.C., Devaux, P., Marie, D., Bergounioux, C., and Petit, P. (1991b) Cytométrie en flux: Application à l'analyse de la ploïdie chez les végétaux. *Biofutur* **105**: 1-14 (Technoscope n47)

Brown, S.C., Renaudin, J.P., Prevot, C., and Guern, J. (1984) Flow cytometry and sorting of plant protoplasts: Technical problems and physiological results from a study of pH and alkaloids in *Catharanthus roseus. Physiologie Végétale* **22**: 541-554

Carayon, P., Bord, A., and Raymond, M. (1991) Simultaneous identification of eight leucocyte subsets of human peripheral blood using three-colour immunofluorescence flow cytometric analysis. *J. Immunol. Methods* **138**: 257-263

Crissman, H.A., and Darzynkiewicz, Z., eds. (1989) "Flow Cytometry", *in:* Methods in Cell Biology, 33, Academic Press, New York

Deslauriers, C., Powell, A.D., Fuchs, K., and Pauls, K.P. (1991) Flow cytometric characterization and sorting of cultured *Brassica napus* microspores. *Biochimica Biophysica Acta* **1091**: 165-172

Dolezel, J. (1991) Flow cytometric analysis of nuclear DNA content in higher plants. *Phytochemical Analysis* **2**: 143-154

Dumain, P., Desnouveaux, R., Bloc'h, L., Leconte, C., Fuhrmann, B., De Colombel, E., Plessis, M-C., and Valery, S. (1990) Use of flow cytometry for yeast and mould detection in process control of fermented milk products: The ChemFlow System - a factory study. *Biotech Forum Europe* **3**: 224-229

Foskett, J.K., and Grinstein, S. (1990) "Noninvasive Techniques in Cell Biology," Wiley-Liss Inc., New York

Galbraith, D.W. (1989) Analysis of higher plants by flow cytometry and cell sorting. *International Review Cytology* **116**: 165-228

Galbraith, D.W., Harkins, K.R., Maddox, J.M., Ayres, N.M., Sharma, D.P., and Firoozabady, E. (1983) Rapid flow cytometric analysis of the cell cycle in intact plant tissues. *Science* **220**: 1049-1051

Gantet, P., Hubac, C., and Brown, S.C. (1990) Flow cytometric fluorescences anisotropy of lipophilic probes in epidermal and mesophyll protoplasts from water-stressed *Lupinus albus* L. *Plant Physiology* **94**: 729-737

Godelle, B., Cartier, D., Marie, D., Brown, S., and Siljak-Yakovlev, S. (1992) Heterochromatin study reveals non-linearity of fluorometry for genomic base composition (unpublished)

Goulas, Y., Moya, I., and Schmuck, G. (1990) Time-resolved spectroscopy of the blue fluorescence of spinach leaves. *Photosynth. Res.* **25**: 299-307

Gray, J., and van der Ploeg, M. (1990) Analytical Cytogenetics. *Cytometry* **11**: 222

Greulich, K.O. (1992) Moving particles by light: No longer science fiction. *Proc. Roy. Microsc. Soc.* **27**: 3-8

Hammatt, N., Lister, A., Blackhall, N.W., Gartland, J., Ghose, T.K., Gilmour, D.M., Power, J.B., Davey, M.R., and Cocking, E.C. (1990) Selection of plant heterokaryons from diverse origins by flow cytometry. *Protoplasma* **154**: 34-44

Harkins, K.R., and Galbraith, D.W. (1987) Factors governing the flow cytometric analysis and sorting of large biological particles. *Cytometry* **8**: 60-70

Horan, P.K., and Slezak, S.E. (1989) Stable cell membrane labelling. *Nature* **340**: 167-168

Jameson, D.M., and Reinhart, G.D. (1989) "Fluorescent Biomolecules: Methodologies and Applications." Plenum Press, New York

June, C.H., and Rabinovitch, P.S. (1988) Flow cytometric measurement of cellular ionized calcium concentration. *Pathol. Immunopatholo. Res.* **7**: 409-432

Le Gall, Y., Brown, S., Marie, D., Mejjad, M., and Kloareg, B. (1992) Production of protoplasts from the red alga *Chondrus crispus*. Application to the quantification of nuclear DNA and to the evaluation of GC%. *Océanis* **18**:11-17

Lister, A. (1989) Flow cytometry for selection of plant cells *in vitro*. *in:* Selection of Plant Cells, Dix, P.J., ed., VCH Verlagsgesellschaft, Weinheim, 39-85

Matine, A. (1990) Biosynthèse d'isoprenoïdes par des protoplastes et plastes isolés du fruit de calamondin. Université de Bordeaux I, Ph.D. thesis

Melamed, M.R., Lindmo, T., and Mendelsohn, M. (1990) "Flow Cytometry and Sorting," 2nd. ed., Wiley-Liss, New York, 824

Mergny, J-L., Sun, J-S., Montenay-Garestier, T., and Helene, C. (1992) Fluorescent-oligodeoxynucleotide conjugates as probes of nucleic acid sequences and structures. *J. Cell Pharmacology*, in press

Perennes, C., Bergounioux, C., and Gadal, P. (1990) Direct assessment of hormone stimulated transcription in protoplasts and isolated nuclei of *Petunia hybrida*. *Plant Cell Reports* **8**: 684-686

Petit, P.X. (1992) Flow cytometric analysis of Rhodamine 123 fluorescence during modulation of the membrane potential in plant mitochondria. *Plant Physiology* **98**: 279-286

Petit, P.X., Diolez, P., and de Kouchkovsky, Y. (1989) Flow cytometric analysis of energy transducing organelles: mitochondria and chloroplasts, *in:* "Flow Cytometry: Advanced research and clinical applications," A. Yen, ed., vol II. CRC Press, Boca Raton Florida, p 271-304

Schröder, W., and Petit, P.X. (1992) Characterisation of chloroplasts, thylakoids and subthylakoid membrane fractions by flow cytometry. *Plant Physiology*, in press

Sigal, L.L., Eversman, S., and Berglund, D. (1988) Isolation of protoplasts from loblolly pine needles and their flow-cytometric analysis for air pollution effects. *Environmental Experimental Botany* **28**: 151-161

Terstappen, L.W., Mickaels, R.A., Dost, R., and Loken, M.R. (1990) Increased light scattering resolution facilitates multidimensional flow cytometric analysis. *Cytometry* **11**: 506-512

Theuillon, F., Raposos, G., Siri, D., and Davoust, J. (1991) Confocal microscopy and subcellular cytometry, *in:* "Confocal Microscopy: Biological Applications", INSERM, Paris, 98-112

Ulrich, I., and Ulrich, W. (1991) High-resolution flow cytometry of nuclear DNA in higher plants. *Protoplasma* **165**: 212-215

Valet, G., and Roth, G. (1991) "Biochemical Cell Functions and Flow Cytometric Data Analysis," Max-Planck-Institut fur Biochemie, Martinsried, Allemagne

Veuskens, J., Marie, D., Hinnisdeals, S., and Brown, S. (1992) Flow cytometry and sorting of plant chromosomes. *in: "*Flow cytometry and Cell Sorting, EMBO Course Methods," A. Radbruch, ed., Springer Monographs: Methods in biology and medicine, Berlin Heidelberg New York, (in press)

Xu, C., Auger, J., and Govindjee, (1990) Chlorophyll a fluorescence measurements of isolated spinach thylakoids obtained by using single-laser-based flow cytometry. *Cytometry* **11**: 349-358

Yentsch, C.M., and Horan, P.K. (1989) Cytometry in the Aquatic Sciences. *Cytometry* **10**: 497-672

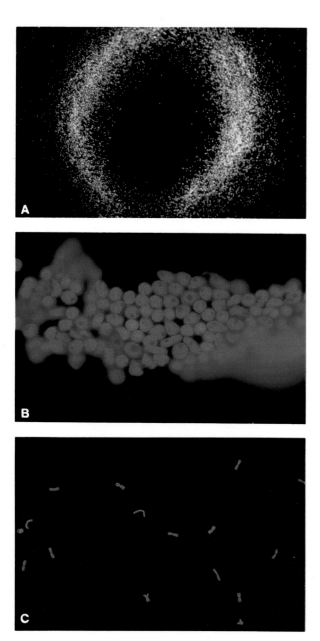

Brown, Group 1: Use of a cell-sorter to separate material according to DNA content after fluorescent staining with bis-benzimide Hoechst 33342.

A. 40,000 G1 nuclei from cultured protoplasts of *Petunia hybrida* sorted onto filter ready for dot hybridization. The regularity of the sorting-droplet deflection onto a 300 × 400 μm zone has produced this perimeter of concentrated nuclei. (Jérôme Conia, Orsay).

B. Sorted G1 nuclei from *Digitaria setigera*. The unfixed nuclei have a diameter of about 9 μm. (Spencer Brown, Gian-Paulo Accotto, Gif-sur-Yvette).

C. Chromosome I of *Petunia hybrida* purified by a sorter (the unfixed chromosome is about 3.6 μm long), (Jérôme Conia, Orsay).

Plate 3

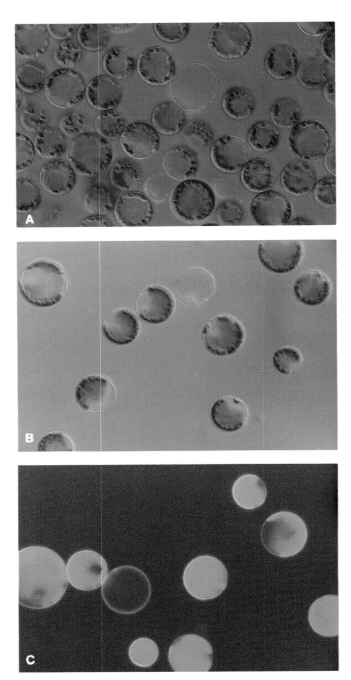

Brown, Group 2: Flow cytometry is a technique to assess cell heterogeneity. Lupin leaf protoplasts labelled with TMA-DPH viewed—(**A,B**): by DIC microscopy; (**C,D**): by epifluorescence microscopy. The red emission derives from chlorophyll, the blue at the plasma membrane from the lipophilic probe whose emission anisotropy reflects membrane fluidity and lipid order. The intense white protoplasts in **D** constitute a subpopulation of autofluorescent cells induced by water-stress. (Pascal Gantet, Camille Hubac, and Spencer Brown, Gif-sur-Yvette).

Plate 4

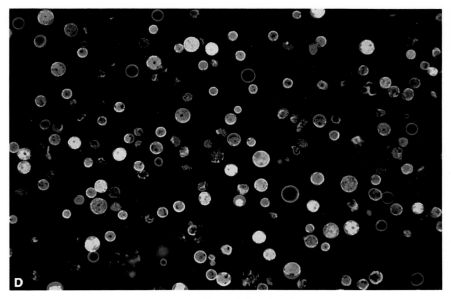

Brown, Group 2 (Continued)

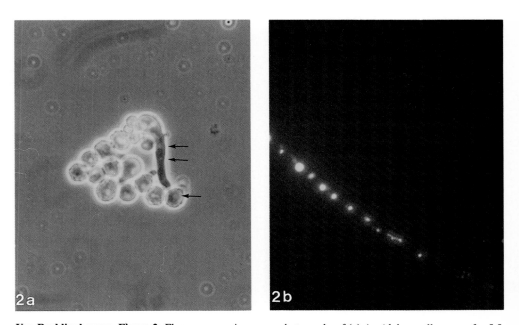

Von Recklinghausen, Figure 2: Fluorescence microscope photographs of (**a**) *A. nidulans* cells grown for 5.5 hours to germ-tube elongation phase of growth in which minimal cell aggregation occurs; (**b**) long hyphae grown for 8 hr in liquid culture. Both figures show dye compartmentation within the fungal cells. (1040 ×)

Plate 5

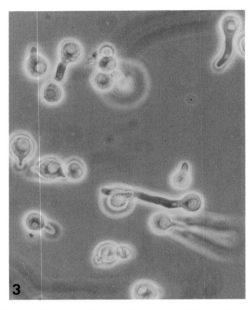

Von Recklinghausen, Figure 3: Dye labelling of *A. nidulans* with pyranine. (1040 ×)

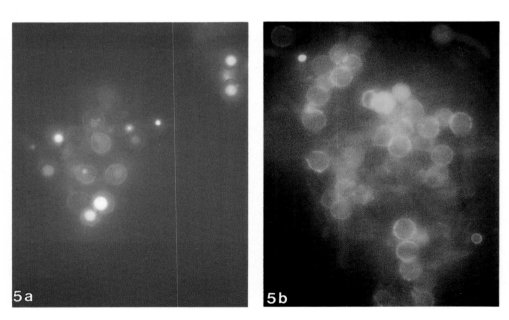

Von Recklinghausen, Figure 5: (a) The fluorescent pattern of chloroquine in *A. nidulans* shows the same label pattern as BCECF/AM. **(b)** The effects of the ionophores nigericin, valinomycin and the protonophore CCCP on BCECF/AM labelling of vacuoles in *A. nidulans*. Due to the dissipation of the H^+ gradient over the vacuolar membranes, no vacuole specific fluorescence can be detected. The dye is seen to be distributed homogeneously in the cytoplasm. (Both 1040 ×)

Plate 6

LASER MICROPERFORATION OF *MEDICAGO SATIVA* ROOT HAIR CELLS

Guenther Leitz[1,2], Armen Kurkdjian[3], Pierre Manigault[3], Abdellah Harim[1], and Karl Otto Greulich[1]

[1]Physikalisch Chemisches Institut, Universität Heidelberg, Im Neuenheimer Feld 253, W-6900 Heidelberg, Germany
[2]Zellenlehre, Fakultät für Biologie, Universität Heidelberg, Im Neuenheimer Feld 230, W-6900 Heidelberg, Germany
[3]C.N.R.S., Institut des Sciences Végétales, Bât. 22, Avenue de la Terrasse, F-91198 Gif-sur-Yvette Cedex, France

ABSTRACT

Microperforation of the plant cell wall of root hairs from *Medicago sativa* can be performed without mechanical contact by using a UV laser microbeam under plasmolytical conditions. Through a gentle and progressive plasmolysis the cells remain viable and the plasma membrane stays intact. The cellular polarity of the root hairs is basically maintained, and it is possible to gain direct access to distinct membrane areas. The plasma membrane is free of visible traces of cell wall fragments, as shown by fluorescence microscopy with Tinopal. The opening in the cell wall (from 1 to 5 μm) leads to a probably turgor-driven movement of protoplasm. The protoplasm can be completely extruded forming a spherical, turgid protoplast which contains the nucleus and shows cytoplasmic streaming. In many cases, only a partial extrusion of the cell content leads to subprotoplasts of various sizes which are still connected to the root hair. Protoplast viability was tested by staining with fluorescein diacetate. The movement of protoplasm depends on the calcium concentration of the extracellular medium. Addition of 0.5 to 1 mM calcium chloride stabilizes the process and clearly increases the yield of successfully opened root hairs. Laser microsurgery avoids the side effects of enzymatic digestion of the plant cell wall. Furthermore, the cell-to-cell interactions through plasmodesmata can be maintained. This preparative laser microtechnique should provide primary material for the study of

Biotechnology Applications of Microinjection, Microscopic Imaging, and Fluorescence, Edited by P.H. Bach *et al.*, Plenum Press, New York, 1993

197

membrane properties, for example, by using the patch clamp technique (Neher, 1992; Sakmann, 1992).

INTRODUCTION

Laser light coupled into a microscope provides a versatile tool for micromanipulation: a laser microbeam (some characteristics of which are summarised in Table 1). For a UV laser microbeam short ultraviolet laser pulses are coupled into a microscope and focused into the object plane, so that microscopic samples in focus can be manipulated (Berns *et al.*, 1981). Thus, a microirradiation with an accuracy of a few hundred nanometers and a time resolution of nanoseconds can be combined with the resolving power of modern light and video microscopy. By focusing the laser light power densities of up to 10^{12} Watt/cm^2 can be reached in the object plane which are sufficient to cut, perforate or fuse microscopic objects without any mechanical contact. Microperforations on the cell surface or in the depth of the cytoplasm with dimensions of less than 1 µm can be performed and precisely aimed through the focusing and scanning mechanisms of the microscope. This technique can be used as an alternative to microinjection by microprojectiles, glass capillaries or chemical substances. Examples for such applications of the laser microbeam are the introduction and expression of DNA in plant cells (Weber *et al.*, 1990) and the micromanipulation of chloroplasts (Weber *et al.*, 1989). Other applications are the fusion of selected pairs of protoplasts (Wiegand *et al.*, 1987) and the microdissection of chromosomes (Monajembashi *et al.*, 1986). The UV laser microbeam can also be combined with a continuous infrared laser (wavelength 1064 nm) which works as a single beam gradient force optical trap (Ashkin *et al.*, 1986). With the optical trap small particles, organelles or even cells can be held and moved relative to their environment (Berns *et al.*, 1989; Block, 1990; Greulich, 1992). In contrast to the UV laser, which causes a selective damage of biological material through absorption, the infrared laser light is only weakly absorbed, thus a considerable nondestructive micromanipulation is possible (Weber *et al.*, 1992). In plant cytology the cell wall often represents a permeability barrier which hinders the micromanipulation of plant cells. Polysaccharides and other compounds in the cell wall behave like traps for nucleic acids, antibodies etc., and the structural strength of the cell wall interferes with the electroporation, microinjection and other techniques in plant cell biology. Therefore, plant protoplasts which are isolated by enzymatic digestion of various tissues to remove the cell walls are used as primary material for experiments (Potrykus and Shillito, 1986; Saunders and Bates, 1987; Zhu and Negrutiu, 1991). Enzymes for protoplast isolation, however, can have deleterious effects on the plasma membrane (Morris *et al.*, 1981; Ishii, 1987; Browse *et al.*, 1988; Hahn and Lörz, 1988; Dugas *et al.*, 1989) and can induce new physiological properties (such as defense

Table 1. Parameters of the lasers.

Leasing medium	Wave-length	Average power	Pulse width	Pulse energy	Pulses per sec.	Remark
Excimer	308nm	10.0W	20ns	100mJ	200	Primary UV source
Dye Laser	340nm	1.5W	20ns	15mJ	200	Tunable wavelength
Nitrogen	337nm	1.0W	5ns	200J	20	Diffraction limited

reactions) in the protoplasts (Chen and Boss, 1990). Therefore, a technique has been developed where the tissues are transferred in a low osmotic medium to induce the release of protoplasts and reduce the incubation time of the plant tissues in the enzyme solution (Elzenga *et al.*, 1991). For methods like patch-clamping which can measure the current flowing through individual ion channels a clean membrane surface is necessary in order to obtain a high resistance seal between the patch pipette and the membrane (Garrill *et al.*, 1992). The membrane surface of protoplasts produced by enzyme treatment is often not "clean" enough for such electrophysiological studies, as is the case, for example, with protoplasts isolated from corn cells in suspension culture (Fairley and Walker, 1989) and from root hairs of *Medicago*. An alternative method to gain access to the plasma membrane is the surgical removal of the cell wall. This technique has been successfully applied to the giant internodal cells from *Chara* (up to 5 cm long), which are refractory to enzyme action (Laver, 1991). Here we use the UV laser microirradiation technique on root hair cells to expose the plant plasma membrane and to induce the formation of subprotoplasts and protoplasts under plasmolytical conditions.

METHODS

The UV Laser Microbeam

The UV laser microbeam consists of an excimer pumped dye laser (Lambda Physik, EMG 102 MSC and FL 2002) or, alternatively, a pulsed ultraviolet nitrogen laser (Laser Science Inc., VSL 337 ND) and an inverted fluorescence microscope (Zeiss, IM 35) equipped with a quartz objective (Zeiss, Ultrafluar 100/NA 1.25 glyc.). The laser beam expanded by a telescope system is directed into the fluorescence illumination path of the microscope and focused to the diffraction limit by the objective lens. The microscopic samples are moved relative to the laser focus by a remote controlled X/Y object stage with 0.25 μm resolution. During irradiation the specimen can be observed either directly through the microscope or with the aid of a TV-camera on a high resolution monitor (Hamamatsu, C1966 and C2130). This system is combined with an image processor (Hamamatsu, C1966 AVEC/VIM).

The wavelength used for microirradiation is 340 nm [p-terphenyl (PTP) used as laser dye] or 337 nm (nitrogen laser). The laser power can be modulated by a variable attenuator. For a detailed description of the laser microbeam system see for example: Wiegand *et al.* (1987); Greulich and Weber (1992); Weber and Greulich (1992).

Fluorescence Microscopy

For cell wall staining, the plants were incubated with the cellulose stain Tinopal (Ciba Geigy, Basel) at a final concentration of 0.5% diluted from a commercial stock solution a few minutes before laser treatment (Falconer and Seagull, 1985). Excitation was at 365 nm and the emitted light was blue (475 nm). Fluorescein diacetate (FDA) was used as a probe for the viability of the protoplasts (Larkin, 1976). The dye was prepared as a 2 mg.ml^{-1} stock solution in acetone and diluted to a final concentration of 0.1 mg.ml^{-1}. Excitation was at 365 nm and the emitted light was green-yellow (450-490 nm).

Plant Material

Seeds of *Medicago sativa* L. (alfalfa) were surface sterilized for 45 minutes using 95% ethanol and then rinsed 5 to 6 times with sterile distilled H_2O. The plants were grown for 48 hours at $21^{\circ}C$ in the dark on 5 mM 2-[N-morpholino]ethanesulfonic acid (MES) buffer adjusted to pH 6.0 with KOH and solidified with 10 g.l^{-1} Bacto Agar (Difco). For the laser experiments, whole plants were transferred on a coverslip in a 150 µl drop of the buffer containing 0.35 M sorbitol or glycine to plasmolyse the root hairs and covered by a second coverslip. Alternatively, the plants were preplasmolysed by gradually increasing the concentration of the osmoticum from 0.2 M via 0.3 M to 0.35 M incubating the plants for 5 minutes at each step. For the growth and viability control, the plants were deplasmolysed by incubating them for a few minutes in a buffer solution containing no osmoticum and then they were deposited for 24 hours on a buffer solution (5 mM MES-TRIS, pH 6.0, containing KCl, NaCl, $CaCl_2$ at 100 µM), solidified with 10 g.l^{-1} Bacto Agar (Difco) and grown in the conditions described above. After 24 hours, roots were checked for root hairs and the lengths of the plants were measured. For the experiments concerning the effect of calcium, the plants were deposited in tissue culture dishes (Petriperm, Bachofer) and submerged in 5 ml medium to prevent evaporation.

RESULTS

With the laser microbeam, the cell wall of *Medicago* root hairs can be perforated while the cells are gently plasmolysed. A few seconds after the opening of the cell wall, the protoplasm swells and tends to fill up the plasmolytic space which forms between the cell wall and the plasma membrane during plasmolysis. This microperforation gives direct access to distinct areas of the plant plasma membrane (for example at the basal or tip region) of the cells. The opening in the cell wall (with a diameter from 1 to 5 µm) leads, in many cases, to an extrusion of protoplasm. The protoplasm is progressively extruded, first a small bud is formed which enlarges and then fills up with the organelles expelled from the root hair cell (Figure 1). The protoplasm behaves like a viscoelastic body which sometimes separates into two parts connected by a small cytoplasmic bridge less than 1 µm in diameter and several µm long (Figure 1 d).

Extrusion of protoplasm can stop at various intermediate stages. It is also possible that the protoplasm remains in the root hair after cell wall perforation. In this case small membrane areas (with a diameter from 1-5 µm), preferentially at the tip of the cell, can be exposed. When the root hairs are rather short (up to 50 µm long) the protoplasm can be released completely forming a spherical, turgid protoplast which contains the nucleus and shows organelle movement. The movement of protoplasm is retarded by increased calcium concentrations. Addition of 0.5 or 1 mM $CaCl_2$ to the medium clearly increases the yield of successfully opened root hairs by slowing down the protoplasmic movement (Table 2). With a higher concentration of 10 mM $CaCl_2$, however, the movement is completely arrested. The viability of the cells after the laser micromanipulation was tested by staining with fluorescein diacetate. The protoplasts showed organelle movement and fluorescein fluorescence after about 5 minutes incubation with fluorescein diacetate. After incubation with Tinopal, the root hairs show a blue fluorescence of their cell wall which is of higher intensity at the tip of the cell. On the contrary, the protoplasts prepared by laser treatment

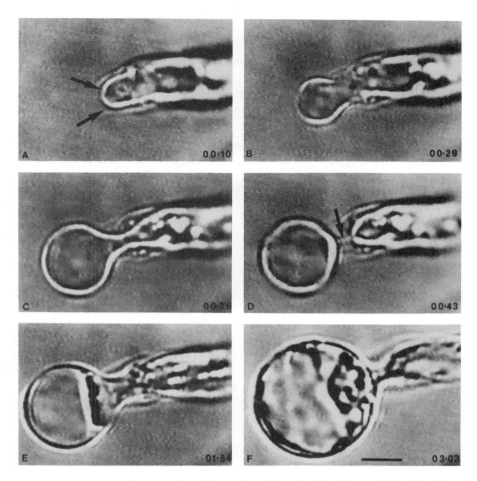

Figure 1. Progressive extrusion of protoplasm from the tip of a root hair cell. (**A**) arrows indicate the laser microperforation in the cell wall. (**D**) small strand of cytoplasm (arrow) connecting the protoplasts. (**A-F**) The time course is indicated in minutes and seconds. Scale Bar = 10 μm.

did not fluoresce, indicating that they are not surrounded by visible traces of cell wall material (Figure 2).

To minimize the effects of plasmolysis and to maintain root hair growth after deplasmolysis, preplasmolysis with gradually increasing glycine concentrations (0.2, 0.3 and 0.35 M for 5 minutes each) was applied. This treatment did not impair root hair development; the root hairs showed cytoplasmic streaming during plasmolysis and were able to grow after deplasmolysis. The roots also grew normally after treatment. The

Table 2. Movement of the protoplasm of *Medicago sativa* root hair cells after laser microperforation of the cell wall.

CaCl$_2$ (mM)	Average speed of movement (μm.sec^{-1})	Lysed protoplasts (percentage)
0	7.2 ± 3.3^b (n = 11)	50 (n = 47)
1.0	0.5 ± 0.1 (n = 17)	15 (n = 64)

^{a} = mean ± SE; n = number of tested root hairs. The speed of the protoplasmic movement (μm.sec^{-1}) was calculated from video recordings of root hairs. Seedlings were incubated in the absence or presence of calcium chloride at the time of the experiments.

difference in length of the roots for treated plants (mean length 3.97 ± 0.43 cm, n = 10) compared to the controls (mean length 4.59 ± 0.53 cm, n = 10) was not significant.

DISCUSSION

Through laser microperforation of the cell wall the plasma membrane of root hair cells can be selectively exposed and the release or partial release of protoplasts can be induced under suitable osmotic and ionic conditions. In contrast to most protoplasts prepared by enzyme digestion, the released protoplasts are free of cell wall fragments detectable by Tinopal fluorescence. Fluorescein accumulation in the cytoplasm after vital staining with fluorescein diacetate indicates esterase activity and membrane integrity. Since protoplasts are osmotically very active and burst rather easily, the correct choice of the plasmolyticum and its concentration are important factors for protoplast isolation (Maheshwari *et al.* 1986). Plasmolysis can create an osmotic stress by decreasing membrane fluidity (Morris *et al.*, 1981; Furtula *et al.*, 1990), modifying the membrane

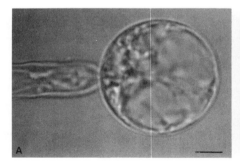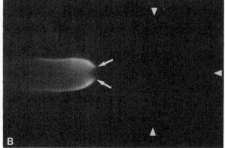

Figure 2. Root hair cell after laser microsurgery. (A). Partial release of protoplasm from the root hair tip. (B) Cell wall staining with Tinopal fluorescence of the same cell. The cell wall of the root hair is fluorescent except at the tip (arrows) which has been perforated by the laser microbeam. The released protoplast (triangles) does not show any fluorescence at all. Scale bar = 10 μm.

potential (Gring *et al.*, 1979) and disturbing the elongation of tip-growing cells (Schnepf *et al.*, 1986). Regular methods of plasmolysing cells for protoplast preparations induce a large decrease of the surface of the plasma membrane and its detachment from the cell wall, so that the polarity of the cells is frequently lost. The situation for plasmolysed root hairs is quite different since these cells preferentially plasmolyse at their tip and the polar axis of the root hairs can remain in existence (Schnepf, 1986). In addition, the plasmolysis applied in our experiment is gentle enough to allow the normal development of root hairs and seedlings after deplasmolysis. The immediate movement of the protoplasm can be interpreted as a turgor-driven movement; also a shift of the wall stress (the elastic pressure of the cell wall) due to the laser microperforation may be involved. The protoplasm was more readily released in young, short root hairs. This corresponds to the fact that the turgor pressure decreases during cell elongation and as a result the osmolarity of young cells is always higher than in old cells (Sakurai, 1991). The requirement for extracellular or cell surface calcium for the successful release of protoplasts can be explained by its stabilizing effect on the plasma membrane (Bangerth, 1979; Brasitus and Dudeja, 1988; Wang *et al.*, 1989). This stabilizing effect probably depends on interactions of the calcium ion with the membrane surface, such as intermolecular bridging of phosphate head groups of membrane phospholipids (Carafoli and Crompton, 1978; Curtain *et al.*, 1988). For enzymatic methods of protoplast isolation calcium ions are also added to enhance the yield and quality of protoplasts (Maheshwari *et al.*, 1986). Since UV laser light can be used for the cell wall perforation with submicrometer accuracy, the orientation of the plant cell can be maintained. Root hairs also provide a favorable model for studying the establishment and maintenance of cell polarity during, for example, the distribution of ion channels in the plasma membrane in relation to polar growth. In addition, when the cell to cell interactions are not disrupted, ion channels can be studied in conditions close to the ones existing *in situ* where plasmodesmata provide pathways for the passage of nutrients, hormones and electrical stimuli in the symplast (Gunning and Overall, 1983; Robards and Lucas, 1990; Clarkson, 1991). The laser microperforation of the cell wall could provide a very attractive method to gain access to the plant plasma membrane for the study of membrane functions in single living cells.

ACKNOWLEDGEMENTS

We would like to thank the DFG (Grant Gr 635/8-1) and the C.N.R.S. for financial support.

REFERENCES

Ashkin, A., Dziedzic, J.M., Bjorkholm, J.E., and Chu, S. (1986) Observation of a single-beam gradient force optical trap for dielectric particles. *Optics Lett.* **11**: 288-290

Bangerth, F. (1979) Calcium-related physiological disorders of plants. *Annu. Rev. Phyto- pathol.* **17**: 97-122

Berns, M.W., Aist, J., Edwards, J., Strahs, K., Girton, J., McNeill, P., Rattner, J. B., Kitzes, M., Hammer-Wilson, M., Liaw, L.-H., Siemens, A., Koonce, M., Peterson, S., Brenner, S., Burt, J., Walter, R., Bryant, P.J., van Dyk, D., Coulombe, J., Cahill, T., and Berns, G.S. (1981) Laser microsurgery in cell and developmental biology. *Science* **213**: 504-513

Berns, M.W., Wright, W.H., Tromberg, B.J., Profeta, G.A., Andrews, J.J., and Walter, R.J. (1989) Use of a laser-induced optical force trap to study chromosome movement on the mitotic spindle. *Proc. Natl. Acad. Sci. USA* **86**: 4539-4543

Block, S.M. (1990) Optical traps: A new tool for biophysics, *in:* "Noninvasive Techniques in Cell Biology", J.K. Foskett, S. Grinstein, eds., Wiley-Liss, New York, 375-402

Brasitus, T.A., and Dudeja, P.K. (1988) Small and large intestinal plasmamembranes: structure and functions, *in:* "Lipid domains and the Relationship to Membrane Function. Advances in Membrane Fluidity, Vol. 2", R.C. Aloia, C.C. Curtain, L.M. Gordon, eds., Alan R. Liss, Inc., New York, 227-254

Browse, J., Somerville, C.R., and Slack, C.R. (1988) Changes in lipid composition during protoplast isolation. *Plant Sci*. **56:** 15-20

Carafoli, E., and Crompton, M. (1978) The regulation of intracellular calcium. *Curr. Top. Membr. Transp.* **10:** 151-216

Chen, Q., and Boss, W.F. (1990) Short-term treatment with cell wall degrading enzymes increases the activity of the inositol phospholipid kinases and the vanadate-sensitive ATPase of carrot cells. *Plant Physiol.* **94:** 1820-1829

Clarkson, D.T. (1991) Root structure and sites of ion uptake, *in:* "Plant roots - the hidden half", Y. Waisel, A. Eshel, and U. Kafkafi, eds., M. Dekker, Inc., New York, 417-453

Curtain, C.C., Gordon, L.M., and Aloia, R.C. (1988) Lipid domains in biological membranes: conceptual development and significance, *in:* "Lipid domains and the Relationship to Membrane Function. Advances in Membrane Fluidity, Vol. 2", R.C. Aloia, C.C. Curtain, L.M. Gordon, eds., Alan R. Liss, Inc., New York, 1-15

Dugas, C.M., Quanning, L., Khan, I.A., and Nothnagel, E.A. (1989) Lateral diffusion in the plasma membrane of maize protoplasts with implications for cell culture. *Planta* **179:** 387-396

Elzenga, J.T.M., Keller, C.P., and Van Volkenburgh, E. (1991) Patch clamping protoplasts from vascular plants - Methods for the quick isolation of protoplasts having a high success rate of gigaseal formation. *Plant Physiol.* **97:** 1573-1575

Fairly, K.A., and Walker, N.A. (1989) Patch clamping corn protoplasts. Gigaseal frequency is not improved by congo red inhibition of cell wall regeneration. *Protoplasma* **153:** 111-116

Falconer, M., and Seagull, R.W. (1985) Immunofluorescent and calcofluor white staining of developing tracheary elements in *Zinnia elegans* L. suspension cultures. *Protoplasma* **125:** 190-198

Furtula, V., Khan, I.A., and Nothnagel, E.A. (1990) Selective osmotic effect on diffusion of plasma membrane lipids in maize protoplasts. *Proc. Natl. Acad. Sci.* **87:** 6532-6536

Garrill, A., Lew, R.R., and Heath, I.B. (1992) Stretch-activated Ca^{2+} and Ca^{2+}-activated K^+ channels in the hyphal tip plasma membrane of the oomycete *Saprolegnia ferax*. *J. Cell Sci.* **101:** 721-730

Gring, H., Polevoy, V., Stahlberg, R., and Stumpe, G. (1979) Depolarization of transmembrane potential of corn and wheat coleoptiles under reduced water potential and after IAA application. *Plant and Cell Physiol.* **20:** 649-656

Greulich, K.O. (1992) Moving particles by light: No longer science fiction. *Proceedings RMS* **27:** 3-8

Greulich, K.O., and Weber, G. (1992) The light microscope on its way from an analytical to a preparative tool. *J. Microscopy* **167:**127-151

Gunning, B.E.S., and Overall, R.L. (1983) Plasmodesmata and cell-to-cell transport in plants. *Bioscience* **33:** 260-265

Hahn, G., and Lörz, H. (1988) Release of phytotoxic factors from plant cell walls during protoplast isolation. *J. Plant Physiol.* **132:** 345-350

Ishii, S. (1987) Generation of active oxygen species during enzymic isolation of protoplasts from oat leaves, *in vitro. Cell Dev. Biol.* **23:** 653-658

Larkin, P.J. (1976) Purification and viability determinations of plant protoplasts. *Planta* **128:** 213-216

Laver, D.R. (1991) A surgical method for accessing the plasma membrane of *Chara australis*. *Protoplasma* **161:** 79-84

Maheshwari, S.C., Gill, R., Maheshwari, N., and Gharyal, P.K. (1986) Isolation and regeneration of protoplasts from higher plants, *in:* "Differentiation of Protoplasts and of transformed Plant Cells", J. Reinert and H. Binding, eds., Springer-Verlag, Berlin, Heidelberg, 3-36

Monajembashi, S., Cremer, C., Cremer, T., Wolfrum, J., and Greulich, K.O. (1986) Microdissection of human chromosomes by a laser microbeam. *Exp. Cell Res.* **167:** 262-265

Morris, P., Linstead, P., and Thain, J.F. (1981) Comparative studies of leaf tissue and isolated protoplasts. *J. Exp. Bot.* **32:** 801-811

Neher, E. (1992) Ion channels for communication between and within cells. *Science* **256:** 498-502

Potrykus, I., and Shillito, R.D. (1986) Protoplasts: Isolation, culture, plant regeneration. *Meth. Enzymol.* **118:** 549-578

Robards, A.W., and Lucas, W.J. (1990) Plasmodesmata. *Annu. Rev. Plant Physiol. Plant Mol. Biol* **41:** 369-419

Sakmann, B. (1992) Elementary steps in synaptic transmission revealed by currents through single ion channels. *Science* **256:** 498-502

Sakurai, N. (1991) Cell wall functions in growth and development - a physical and chemical point of view. *Bot. Mag. Tokyo* **104:** 235-251

Saunders, J.A., and Bates, G.W. (1987) Chemically induced fusion of plant protoplasts, *in:* "Cell Fusion", A.E. Sowers, ed., Plenum Press, New York, 497-520

Schnepf, E., Deichgräber, G., Bopp, M. (1986) Growth, cell wall formation and differentiation in the protonema of the moss: *Funaria hygrometrica:* effect of plasmolysis on the developmental program and its expression. *Protoplasma* **133:** 50-65

Schnepf, E. (1986) Cellular polarity. *Annu. Rev. Plant Physiol.* **37:** 23-47

Wang, H., Cutler, A.J., Saleem, M., and Fowke, L. C. (1989) Microtubules in maize protoplasts derived from cell suspension cultures: effect of calcium and magnesium ions. *Eur. J. Cell Biol.* **49:** 80-86

Weber, G., Monajembashi, S., Greulich, K.O., and Wolfrum, J. (1989) Uptake of DNA in chloroplasts of *Brassica napus* (L.) facilitated by a UV-laser microbeam. *Eur.J.Cell Biol.* **49:** 73-79

Weber, G., Monajembashi, S., Wolfrum, J., and Greulich, K.O. (1990) Genetic changes induced in higher plant cells by a laser microbeam. *Physiol. Plant.* **79:** 190-193

Weber, G., and Greulich, K.O. (1992) Manipulation of cells, organelles, and genomes by laser microbeam and optical trap. *Int. Rev. Cytol.* **133:** 1-41

Weber, G., Leitz, G., Seeger, S., and Greulich, K.O. (1992) Laser microbeams as optical tools for living cells. *J. Exp. Bot.* **43:** 250: 72

Wiegand (Steubing), R., Weber, G., Zimmermann, K., Monajembashi, S., Wolfrum, J., and Greulich, K.O. (1987) Laser induced fusion of mammalian cells and plant protoplasts. *J. Cell Sci.* **88:** 145-149

Zhu, X.-Y., and Negrutiu, I. (1991) Isolation and culture of protoplasts, *in:* "A Laboratory Guide for Cellular and Molecular Plant Biology", I. Negrutiu, G. Gharti-Chhetri, eds., Birkhäuser, Basel, 18-27

APPLICATION OF pH-SENSITIVE FLUORESCENT
DYES TO FUNGI

I.R. von Recklinghausen[1], D. Molenaar[2], W.N. Konings[2], and
J. Visser[1]

[1]Dept. of Genetics, Section of Molecular Genetics,
Wageningen Agricultural University, Dreijenlaan 2, 6703 HA
Wageningen, The Netherlands
[2] Dept. of Microbiology, University of Groningen, Kerklaan
30, 9751 NN Haren, The Netherlands

SUMMARY

The use of the probes 2',7'-bis-(2-carboxyethyl)-5(and-6)-carboxy-fluorescein
acetoxymethyl ester and pyranine to measure the internal pH of a suspension of *Aspergillus
nidulans* (Ascomycetes) germlings by spectrofluorimetry was investigated. Growth
conditions were established, by which it was possible to keep the cells in a homogeneous
suspension, thus allowing measurements in cuvettes.

A microscopic investigation of the distribution of the pH-sensitive probes
surprisingly revealed that the dyes tested do not label the cytoplasm but are accumulated
preferentially in a cell compartment of variable size, dependent on the location within the
fungal cell. In transmission electron microscopic studies these organelles have been
identified as vacuoles, the sizes of which become reduced when situated away from the
spore head towards the hyphal tip. Additionally, specific staining with the dye chloroquine
and the loss of the bright fluorescence in these vacuoles by addition of the protonophore
carbonylcyanide-m-chloro-phenylhydrazone confirms that the probe is located in vacuoles
and not in nuclei, which may have the same size. The response of the vacuoles to several
ionophores and protonophores and to the external application of NH4Cl has been tested.
The results show that fluorescence microscopy and spectrofluorimetry are suitable tools to
study the complicated pH-regulation mechanisms and subcellular distribution of protons in
Aspergillus species.

Biotechnology Applications of Microinjection, Microscopic Imaging, and
Fluorescence, Edited by P.H. Bach *et al.*, Plenum Press, New York, 1993

207

INTRODUCTION

In the last few years several fluorescent pH-indicator probes have been developed (Bassnet *et al.*, 1990; Casadio, 1991; Haugland, 1989; Buckler and Vaughan-Jones, 1990; Graber *et al.*, 1986; Damiano *et al.*, 1984) and rapid progress has been made in employing these dyes in techniques, such as digital imaging microscopy (Cole *et al.*, 1990; Grapengiesser *et al.*, 1989; Maxfield, 1989; Montrose *et al.*, 1987; Seksek *et al.*, 1991) and spectrofluorimetry (Luvisetto *et al.*, 1991a, b; Rahim *et al.*, 1991; Tsujimoto *et al.*, 1988) in order to measure the intracellular pH in various cell types. However, very little is known about distribution mechanisms, i.e. dye accumulation and efflux (Horobin and Rashid, 1990; Rashid *et al.*, 1991; Thomas *et al.*, 1979), and modifications of probe characteristics (Graber *et al.*, 1986; Seksek *et al.*, 1991) inside the cells due to binding to proteins and fluorescence quenching. These parameters have to be carefully tested for each dye and for each cell type.

Determination of the internal pH of fungi has proven to be a difficult problem. Most pH regulation studies in eukaryotic, non-epithelial cells have been performed with yeasts or the fungus *Neurospora* utilizing methods such as ^{15}N- or ^{31}P-NMR spectroscopy (Gillies *et al.*, 1982; Campbell-Burk and Shulman, 1987; Greenfield et al., 1987; Pilatus and Techel, 1991; Legerton *et al.*, 1983; Nicolay *et al.*, 1982), microelectrodes (Kotyk, 1989; Slayman and Zuckier, 1989), the distribution of weak acids (Gresík *et al.*, 1991) and to some extent pH-sensitive probes (Roos and Slavík, 1987; Slavík, 1982; Slavík, 1983).

So far, the application of pH measurements with fluorescent dyes to hyphal fungi has been limited to the marine fungus *Dendryphiella salina* (Davies *et al.*, 1990). Spectrofluorimetry and pH-sensitive fluoroprobes as a very fast, sensitive and non-invasive technique have been chosen to measure the intracellular pH in intact *Aspergillus* cells. In order to study under physiological conditions the various ion transport and bioenergetic systems which affect proton gradients, the fluorescent dyes pyranine and 2′, 7′-bis-(2-carboxyethyl)-5(and-6)-carboxyfluorescein acetoxymethyl ester (BCECF/AM) have been tested. *Aspergillus nidulans* serves as a good model system because of its capacity to grow over a wide range of external pH, from 2.5 to 9 (Caddick *et al.*, 1986), and its ability to change actively the pH of its surroundings, e.g. by acidifying the medium via excretion of acids.

In this report some technical aspects concerning cell-growth and dye-loading will be briefly introduced. Subsequently, some aspects on probe distribution and characteristics of the cell compartment in which the dye accumulates are given and, finally, potential uses of pH-sensitive dyes to monitor pH changes are discussed. We expect that this method would be suitable for studying the dynamic changes of vacuolar pH in response to changes in metabolism and environmental conditions.

MATERIALS AND METHODS

Abbreviations

BCECF/AM = 2′,7′-bis-(2-carboxyethyl)-5(and-6)-carboxyfluorescein acetoxy-methyl ester; CCCP = carbonylcyanide-m-chloro-phenylhydrazone; DMSO, dimethyl sulfoxide; HEPES = N-2-hydroxy-ethylpiperazine-N′-2-ethanesulfonic acid; MES = 2-(N-morpholino)ethanesulfonic acid; pH_i = intracellular pH; pyranine = 8-hydroxy- 1,3,6-pyrenetrisulfonic acid.

Chemicals and Reagents

BCECF/AM, pyranine and nigericin were purchased from Molecular Probes Inc., Eugene, OR., USA. Stock solutions (10 mM) of the fluorescent dyes in DMSO were kept in the dark and stored at -20 °C. Other chemicals were purchased from Sigma Chemical Co., St. Louis, MO. (valinomycin, chloroquine) and Aldrich, Vlaxel, NL (CCCP, monensin).

Strain and Culture Conditions

For the preparation of conidiospores *Aspergillus nidulans* WGO96 strain (yA2, pabaA1) was grown for 5 days at 37 °C on agar plates using complete medium with 15 mM sucrose as carbon source (Pontecorvo *et al.,* 1953). For germination, 10^7 spores/ml were inoculated in liquid minimal medium (Pontecorvo *et al.,* 1953) supplemented with 2 μg/ml p-aminobenzoic acid, appropriate vitamins and 30 mM sucrose as carbon source. After growth for 5.5 hours at 37 °C in a New Brunswick Scientific orbital shaker (200 rpm), the young mycelium was harvested by filtration using micro sieves (Endecotts, London), washed with cold saline (0.85% NaCl) and resuspended in media containing, per litre: 6 g $NaNO_3$, 1.5 g KH_2PO_4, 0.5 g KCl, 0.5 g $MgSO_4.7 H_2O$, 3.9 g MES, 4.7 g HEPES plus trace elements and 4-aminobenzoic acid (pH 7.20). The time point of harvesting is critical, as pointed out in "Results", and was determined by microscopic inspection.

Loading of Cells with BCECF/AM and Pyranine

The medium above mentioned was also used for loading the fungi with the pH-sensitive dyes and for further spectrofluorimetric studies. Cells were loaded with BCECF by exposure for 30 to 40 minutes to 10 μM of the acetoxymethylester at 25 °C in the dark. The acetoxymethylester of BCECF has capped carboxyl groups and is an uncharged, membrane-permeable molecule, which appparently can enter intact cells. The permeable precursor is then cleaved by internal cell esterases to yield the charged, pH-sensitive dye which remains trapped inside (Rink *et al.,* 1982).

The membrane-impermeable pH-indicator pyranine was loaded into fungal cells by application of the acid-shock method recently developed for *Lactococcus lactis* and consists of briefly treating a dense cell suspension with acid in the presence of the probe (Molenaar *et al.,* 1991). Extracellular dye was washed out with the previously mentioned buffer (pH 7.20). The cells were then concentrated and resuspended in 1 ml of buffer and kept on ice until use.

Dye-loading was monitored at the microscopic level with a Zeiss Epi-Fluorescence Axioskop using a filter combination giving excitation between 400-440 nm. For photographs a 400 Iso Ektachrome slide film was used.

Spectrofluorimetry

Fluorescence spectra were recorded on a Perkin-Elmer LS-50 Luminescence Spectrometer, equipped with a stirred, thermostatted cell holder set to 30 °C. The cuvette contained 3 ml of the desired buffer and 60 to 80 μl probe-loaded cells, depending on the concentration of BCECF. The dye was excited at 502 nm and the emission recorded at 525 nm. A slit width of 10 nm was used for both. The fluorescence signal was averaged over time intervals of 1 second.

For correlating the measured fluorescence signal to intracellular pH, the probe was calibrated intracellularly using nigericin, valinomycin and CCCP in calibration-buffers of different pH values containing high amounts of potassium, in order to dissipate proton gradients across membranes.

The emission measured at 502 nm excitation and the fluorescence intensity after Triton-X-100 permeabilization and alkalinisation to pH 11-12 was used for the calculation of the fluorescence ratio R [in this pH range fluorescence is high and almost pH insensitive (Molenaar *et al.,* 1991)]. A calibration curve (fluorescence ratio as a function of pH) was constructed from these measurements.

The pH changes of the external buffer of the cell suspension were followed with a Consort P944 pH meter.

Electron Microscopy

For studies of cell morphology, *A. nidulans* WG096 spores were germinated and harvested as described for the fluorescence experiments. The cells were fixed in aqueous 2.5% (w/v) $KMnO_4$ at room temperature for 30 minutes, washed 3 times in water and post-stained with 1% uranylacetate overnight. The dehydration of the samples in a graded ethanol series was performed in a Biorad H 2500 Microwave processor at a temperature of $25^{o}C$. Subsequently the samples were embedded in Epon 812.

Ultra thin sections cut with a diamond knife were examined in a Philips EM 300 electron microscope operating at 60 kV.

RESULTS

Growth Conditions

Due to the difficulty in obtaining homogeneous cell suspensions of *Aspergillus* mycelium (i.e., unclumped cells) when grown in liquid medium, it was necessary to find growth conditions in which no cell aggregation occurred in order to be able to measure the intracellular pH of a cell population in the fluorimeter.

Growth experiments (Figure 1) showed that young spores (germlings) of *A. nidulans* which have been germinated for 5.5 hours showed minimal aggregation and thus were suitable to work with in suspension. The time point of harvesting is critical, because the spores are in a special, short germination phase which corresponds to the germ-tube elongation stage classified by Campbell *et al.* (1971) and Florance *et al.* (1972). Later developmental stages are unsuitable for the preparation of homogeneous suspensions because branching of the hyphae yields an additional factor leading to aggregate formation.

Loading with BCECF/AM and Pyranine

In order to choose a suitable fluorescent pH indicator two fluorescent dyes, namely pyranine (8-hydroxy 1,3,6-pyrenetrisulfonate) and BCECF/AM (2′,7′-bis-(2-carboxyethyl)-5(and-6)-carboxyfluorescein acetoxymethyl ester) were tested and the intracellular distribution of the dyes monitored microscopically.

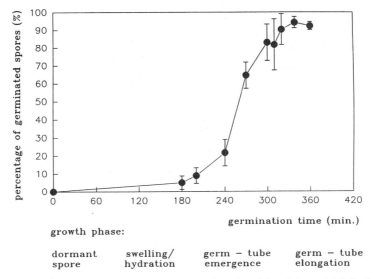

growth phase:

| dormant
spore | swelling/
hydration | germ – tube
emergence | germ – tube
elongation |

Figure 1. Time course of conidial germination (see Materials and Methods for growth conditions) and classification of growth phases as judged by light microscopic investigations. Harvesting time of the fungi is critical and must be done prior to branching of the hyphae in order to obtain suspensions of nonaggregated cells.

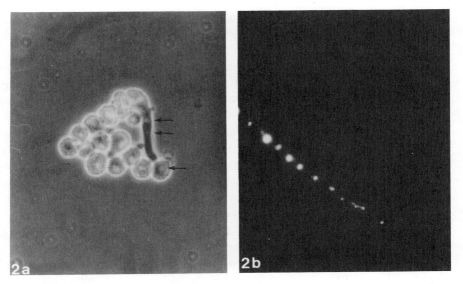

Figure 2. Fluorescence microscope photographs of: **a)** *A. nidulans* cells grown for 5.5 hours to germ-tube elongation phase of growth in which minimal cell aggregation occurs; **b)** long hyphae grown for 8 hours in liquid culture. Both figures show dye compartmentation within the fungal cells (1040 x). See colour plate 5.

These investigations showed that BCECF/AM does not seem to label the cytoplasm of young spores (Figure 2a,b) but is preferentially accumulated in a cell compartment of variable size, the latter property being dependent on its location within the fungal cell (Figure 2).

Bright BCECF/AM fluorescence can be located in one or more organelles within the sporehead and in somewhat smaller cell compartments within the hyphae (Figure 2a arrows), suggesting that the dye is concentrated there by an unknown mechanism. It can be clearly seen from Figure 2b, that the size of the cell compartments stained by fluorescence reduces in size when situated away from the spore head towards the hyphal tip. This labelling pattern strongly indicates that the probe is trapped in the vacuolar compartments of the fungi. It is well-known that the vacuolar size varies within the cell (Bracker, 1967).

Fluorescence of BCECF trapped in mitochondria, as described for other cell types (Luvisetto *et al.,* 1991a,b) could not be detected in young spores of *A. nidulans*, although these organelles are present in high numbers within these fungal cells.

In contrast to the BCECF-labelling, the fluorescence exhibited by pyranine loaded fungi seems to be more heterogeneous. This is due to the presence, within the preparation, of cells in which, a) no fluorescence is detected, b) the fluorescence is located in the cytoplasm (Figure 3, upper right corner), and c) a fluorescence pattern similar to that observed in BCECF/AM loaded cells is found. The uneven distribution of the dye, and a reduced viability of pyranine incubated cells in comparison to BCECF/AM labelled cells may be caused by the acid shock applied during the loading process. Due to the unpredictable distribution of pyranine, further experiments were exclusively performed with the indicator dye BCECF/AM.

Identification of the Labelled Cell Compartment

Ultrastructural transmission electron microscopic studies and light microscopic investigations (see above) revealed that the organelles in which the pH indicator dyes accumulate are vacuoles and vesicles.

A typical *A. nidulans* cell in longitudinal section, which has been grown for 5.5 hours, is shown in Figure 4a. Usually one large, or a few smaller, electron transparent vacuoles (Va, Figure 4a) are found in the area opposite the germ-tube. This location corresponds to the site of acccumulation of the indicator dye in the fluorescence pictures.

The vacuoles (Va) in the hyphae are somewhat smaller due to the reduced diameter of the hyphae compared to the sporeheads (Figure 4a). Some vacuoles contain multivesicular bodies and organelles, such as mitochondria (M, Figure 4b, arrow), and obviously perform degradation tasks. Beneath the vacuoles, one or two nuclei are located in close relation to the endoplasmic reticulum and are surrounded by a large number of mitochondria (M, Figure 4b), which indicates high metabolic activity.

In order to further verify that BCECF/AM is indeed accumulated in an acidic vacuolar compartment and not in the cell nucleus, a specific fluorescent marker for acidic cell organelles was introduced for comparison of the fluorescence patterns. *A. nidulans* cells were loaded with chloroquine in the same way as BCECF/AM. Chloroquine, which is known to be concentrated in yeast vacuoles (Lenz and Holzer, 1984) and mammalian lysosomes (de Duve *et al.*, 1974), requires excitation at 390 nm to monitor the fluorescence (Figure 5a). The label pattern corresponds to that exhibited with BCECF/AM and thus the same acidic vacuolar cell compartments are stained (fluorescence patterns of Figure 2 and Figure 5a).

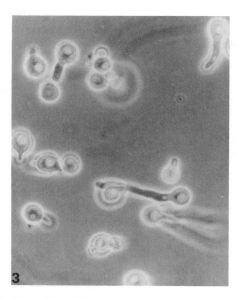

Figure 3. Dye labelling of *A. nidulans* with pyranine (1040 x). See colour plate 6.

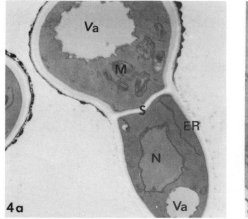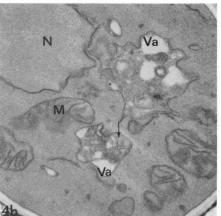

Figure 4. Typical appearance of *A. nidulans* germlings (longitudinal sections) grown for 5.5 hours. Transmission electron microscopy and light microscopy identify the organelles in which the pH-sensitive dyes accumulate as vacuoles and vesicles. ER: endoplasmic reticulum, M: mitochondrion, N: nucleus, S: septum, Va: vacuole (**a**) 8700 x, (**b**) 23600 x.

Perturbation of Proton Gradients

The distribution of BCECF/AM in *A. nidulans* germlings suggested the possibility of measuring pH and pH changes at the subcellular level. Ionophores and protonophores, which equilibrate proton gradients over membranes (Pressman, 1976; Ohkuma and Poole, 1978), are usually used for the preparation of calibration curves in order to correlate the measured fluorescence signals to pH values (fluorescence ratio as a function of pH).

The external application of either monensin (20 µM) or valinomycin (20 µM) to BCECF/AM loaded *A. nidulans* germlings showed very little effect on vacuolar membranes, as determined from negligible changes in fluorescence within the vacuolar compartment as detected using both spectrofluorimetry and microscopy.

The addition of the ionophore nigericin (20 µM) or the protonophore CCCP (30 µM) led to a more pronounced change in fluorescence level but not to a total dissipation of proton gradients and thus to incomplete equilibration. The combination of nigericin, valinomycin and CCCP was effective as indicated by the loss of bright fluorescence in the vacuole, suggesting pH change and breakdown of proton gradient between vacuole and cytoplasm (Figure 5b). Under these conditions the collapse of the proton gradient was additionally accompanied by a release of fluorescent probe into the cytoplasm, which can be concluded from the homogeneous distribution of fluorescence within the cells (Figure 5b).

The vacuolar pH could also be perturbed by the addition of 10 mM NH4Cl to the external medium (pH 7.08), which led to an alkalinisation of the vacuolar compartment by 0.22 ± 0.03 pH-units (n = 4) as shown in Figure 6.

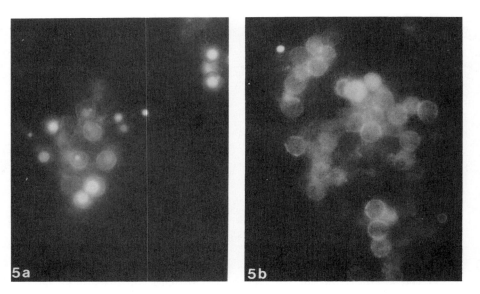

Figure 5. (a) The fluorescence pattern of chloroquine in *A. nidulans* shows the same label pattern as BCECF/AM (1040 x). (b) The effects of the ionophores nigericin, valinomycin and the protonophore CCCP on BCECF/AM labelling of vacuoles in *A. nidulans*. Due to the dissipation of the H^+ gradient over the vacuolar membranes, no vacuole specific fluorescence can be detected. The dye is also seen to be distributed homogeneously in the cytoplasm (1040 x). See also colour plate 6.

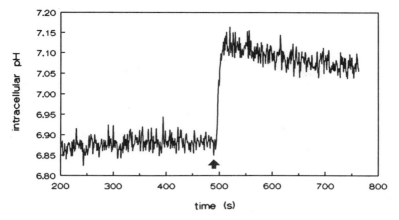

Figure 6. Vacuolar steady state pH and addition of 10 mM NH4Cl (vertical arrow) to the external medium (pH 7.08). NH4Cl application leads to an alkalinisation of the vacuolar compartment of 0.22 ± 0.03 pH-units (n = 4). Thus, pH-sensitive fluorescent dyes can be used to monitor selectively pH changes of the vacuolar compartment of the fungal cells.

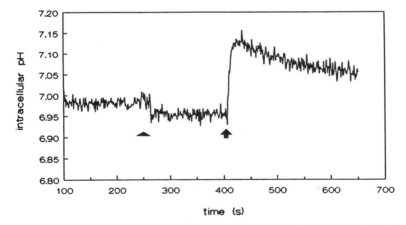

Figure 7. The addition of external glucose (arrow head) results in an acidification of the vacuoles by 0.07 ± 0.01 pH units (n = 5). Subsequent addition of 10 mM NH4Cl (vertical arrow) to the external medium (in the presence of 1% glucose as carbon source) leads to an alkalinisation.

The uptake of easily utilisable carbon sources, such as glucose, resulted in a temporary acidification of the vacuoles by 0.07 ± 0.01 pH-units (n = 5). A subsequent NH4Cl addition (10 mM) had the same effect as noted above (Figure 7).

Thus, pH-sensitive fluorescent dyes can be used to monitor selectively the intracellular pH and pH changes of the vacuolar compartment of the fungal cells.

CONCLUSIONS

The studies presented above demonstrate that spectrofluorimetry and pH-sensitive dyes, such as BCECF/AM, can be used to monitor pH changes in *A. nidulans* germlings at the subcellular level and that it is possible to study the involvement and contribution of the vacuolar compartment in the complex pH regulation processes in these fungi. The function of the latter in cellular acid-base balance is, however, currently ill defined.

Sulfonic acid derivates, such as pyranine are relatively difficult to load into cells. Usually they must be microinjected, scrape-loaded, loaded by endocytic processes or temporary membrane disruption, or they must be studied by patch clamp techniques. Loading *A. nidulans* germlings with pyranine using the acid-shock method (Molenaar *et al.*, 1991) led to inhomogeneous distribution of the fluorescence. Thus it is not possible to measure the intracellular pH specifically with spectrofluorimetry. The applicability of this method for loading pyranine into *A. nidulans* seems to be further limited because the temporary acid shock during the loading procedure seems to reduce viability of the cells, although the fungus is known to be able to grow over a wide pH range from pH 2.5 to 9 (Caddick *et al.*, 1986).

Loading of *A. nidulans* germlings with the pH-sensitive probe BCECF/AM showed that the indicator was concentrated in cell compartments located in the spore head and in the short hyphae of the fungi. This phenomenon indicates that loading of cells with the highly esterified form of a fluoroprobe is not a routine method but that the distribution of the dye has to be monitored carefully and that the biological behaviour can vary strongly, depending on the cell type used. The location of fluorescein derivatives within different cells is considered to be mainly cytoplasmic (Grapengiesser *et al.*, 1989) but accumulation in mitochondria (Luvisetto *et al.*, 1991a,b) and acidic organelles (Ohkuma and Poole, 1978; Geisow and Evans, 1984; Geisow *et al.*, 1984) have been reported in the past.

The nature of the organelles containing the fluoroprobe in *A. nidulans* was identified as the vacuolar compartment because:-

a) the fluorescence pattern of BCECF/AM corresponds to that exhibited by chloroquine, a specific fluorescent marker for labelling acidic cell compartments, such as vacuoles and lysosomes (Lenz and Holzer, 1984; de Duve *et al.*, 1974),

b) application of nigericin enhances fluorescence intensity, which corresponds to an alkalinisation. It is well known from the literature that acidic potassium ionophores cause the pH of lysosomes to increase (Ohkuma and Poole, 1978),

c) transmission electron microscopy studies showed that the vacuolar size depends on the location within the fungal cell, i.e. the size reduces away from the spore head towards the hyphal tip. The same can be seen in the fluorescence pattern of the BCECF labelled organelles.

The concentration mechanism of BCECF/AM is unknown but might be partially due to the existing proton gradients between the cell compartments (i.e. vacuole and cytoplasm) because addition of the protonophore CCCP does not only lead to a breakdown of the proton gradient but also to a release of dye into the cytoplasm. BCECF/AM accumulation might also be a result of high vacuolar esterase activities which are present in high amounts in fungal vacuoles (Klionsky *et al.*, 1990). The strong fluorescence signal exhibited by the labelled organelles suggests a very efficient concentration mechanism for the dye because the vacuolar interior would be acidic enough (around pH 6.0) to quench the fluorescence signal.

It is desirable to obtain calibration curves for the pH indicator when it is in an environment very similar to that found in living cells (Maxfield, 1989; Roos and Boron, 1981). For this reason titration curves are prepared in the presence of labelled cells incubated with ionophores and protonophores, which are assumed to generate conditions which equalize cytoplasmic and external pH (Pressman, 1976). Tests of the effects of several ionophores in *A. nidulans* revealed that the ionophores on their own only led to an incomplete proton release out of the vacuoles, but a combination of several ionophores and a protonophore (CCCP, nigericin and valinomycin) was effective in dissipating the proton gradients across the vacuolar membranes. The experiments showed a response of the vacuolar membranes on nigericin and CCCP addition but their effect on the plasma membrane is yet unclear. These results suggest that for fungi it is necessary to describe the action of the proton dissipating substances in more detail, because it has to be carefully evaluated:-

a) whether the chemicals can reach internal membranes,

b) on which membrane they act efficiently.

It has been reported, for example, that valinomycin and nigericin act as ionophores in the inner mitochondrial membrane but not in the plasma membrane of intact yeast cells (Kovác et al., 1982a,b). In contrast, the plasma membrane of *Trichoderma viride* responds to valinomycin (Gresík et al., 1991). Nigericin is reported to dissipate proton gradients in the marine fungus *Dendryphiella salina* (Davies et al., 1990). The action of ionophores and protonophores, which are mainly metabolites produced by fungi (Pressman, 1976), on *A. nidulans* membranes are currently under investigation.

This approach is very promising for studying the complex pH regulation mechanisms in *A. nidulans*. This conclusion is based on the fact that the results presented here show that the vacuolar cell compartment has an important function in pH regulation. Externally applied bases induce pH transitions within the organelle and it appears that this compartment also helps to diminish cytoplasmic pH changes.

ACKNOWLEDGEMENTS

We are grateful to Dr. M. Veenhuis (Laboratory of Electron Microscopy, Biological Centre, University of Groningen, NL) for valuable information about preparing spores for electron microscopy and for the use of equipment and facilities. The skilful technical assistance of Klaas Sjollema is gratefully acknowledged.

This study was financially supported by the E.E.C. Bridge program PL890136.

REFERENCES

Bassnett, S., Reinisch, L., and Beebe, D.C. (1990) Intracellular pH measurement using single excitation-dual emission fluorescence ratios. *Am. J. Physiol.* **258** (*Cell Physiol.* **27**): 171-178

Bracker, C.E. (1967) Ultrastructure of fungi. *Ann. Rev. Phytopathol.* **5**: 343-374

Buckler, K.J., and Vaughan-Jones, R.D. (1990) Application of a new pH-sensitive fluoroprobe (carboxy-SNARF-1) for intracellular pH measurement in small, isolated cells. *Pflügers Arch.* **417**: 234-239

Caddick, M.X., Brownlee, A.G., and Arst, H.N. (1986) Regulation of gene expression by pH of the growth medium in *Aspergillus nidulans*. *Mol. Gen. Genet.* **203**: 346-353

Campbell, C.K. (1971) Fine structure and physiology of conidial germination in *Aspergillus fumigatus*. *Trans. Br. Mycol. Soc.* **57**: 393-402

Campbell-Burk, S.L., and Shulman, R.G. (1987) High-resolution NMR studies of *Saccharomyces cerevisiae*. *Ann. Rev. Microbiol.* **41**: 595-616

Casadio, R. (1991) Measurements of transmembrane pH differences of low extents in bacterial chromatophores. A study with the fluorescent probe 9-amino,6-chloro,2-methoxyacridine. *Eur. Biophys. J.* **19**: 189-201

Cole, L., Coleman, J., Evans, D., and Hawes, C. (1990) Internalisation of fluorescein isothiocyanate and fluorescein isothiocyanate-dextran by suspension-cultured plant cells. *J. Cell. Sc.* **96**: 721-730

Damiano, E., Bassilana, Rigaud, J.-L., and Leblanc, G. (1984) Use of the pH-sensitive fluorescence probe pyranine to monitor internal pH changes in *Escherichia coli* membrane vesicles. *FEBS Lett.* **166**:120-124

Davies, J.M., Brownlee, C., and Jennings, D.H. (1990) Measurement of intracellular pH in fungal hyphae using BCECF and digital imaging microscopy. Evidence for a primary proton pump in the plasmalemma of a marine fungus. *J. Cell Sc.* **96**: 731-736

de Duve, C., de Barsy, T., Poole, B., Trouet, A., Tulkens, P., and van Hoof, F. (1974) Lysosomotropic agents. *Biochem. Pharmacol.* **24**: 2495-2531

Florance, E.R., Denison, W.C., and Allen, T.C. (1972) Ultrastructure of dormant and germinating conidia of *Aspergillus nidulans*. *Mycologia* **61**: 115-122

Geisow, M.J. (1984) Fluorescein conjugates as indicators of subcellular pH. A critical evaluation. *Exp. Cell Res.* **150**: 29-35

Geisow, M.J., and Evans, W.H. (1984) pH in the endosome. Measurements during pinocytosis and receptor-mediated endocytosis. *Exp. Cell Res.* **150**: 36-46

Gillies, R.J., Alger, J.R., den Hollander, J.A., and Shulman, R.G. (1982) Intracellular pH measured by NMR: Methods and results, *in:* "Intracellular pH: Its measurement, regulation, and utilization in cellular functions", Alan R. Liss, Inc. New York, 79-104

Graber, M.L., DiLillo, D.C., Friedman, B.L., and Pastoriza-Munoz, E. (1986) Characteristics of fluoroprobes for measuring intracellular pH. *Anal. Biochem.* **156**: 202-212

Grapengiesser, E., Gylfe, E., and Hellman, B. (1989) Regulation of pH in individual pancreatic ß-cells as evaluated by fluorescence ratio microscopy. *Biochim. Biophys. Acta* **1014**: 219-224

Greenfield, N.J., Hussain, M., and Lenard, J. (1987) Effects of growth state and amines on cytoplasmic and vacuolar pH, phosphate and polyphosphate levels in *Saccharomyces cerevisiae*: a ^{31}P-nuclear magnetic resonance study. *Biochim. Biophys. Acta* **926**: 205-214

Gresík, M., Kolarova, N., and Farkas, V. (1991) Hyperpolarization and intracellular acidification in *Trichoderma viride* as a response to illumination. J. Gen. Microbiol. 137: 2605-2609

Haugland, R.P. (1989) "Handbook of fluorescent probes and research chemicals", Molec. Probes, OR., USA, 86-94

Horobin, R.W., and Rashid, F. (1990) Interactions of molecular probes with living cells and tissues. 1. Some general mechanistic proposals, making use of a simplistic chinese box model. *Histochem.* **94**: 205-209

Klionsky, D.J., Herman, P.K., and Scott D.E. (1990) The fungal vacuole: composition, function, and biogenesis. *Microbiol. Rev.* **54**: 266-292

Kotyk, A. (1989) Proton extrusion in yeast. *Meth. Enzymol.* **174**: 592-603

Kovác, L., Böhmerová, E., and Butko, P. (1982a) Ionophores and intact cells. I. Valinomycin and nigericin act preferentially on mitochondria and not on the plasma membrane of *Saccharomyces cerevisiae*. *Biochim. Biophys. Acta* **721**: 341-348

Kovác, L., Poliachová, V., and Horváth, I. (1982b) Ionophores and intact cells. II. Oleficin acts on mitochondria and induces disintegration of the mitochondrial genome in yeast *Saccharomyces cerevisiae*. *Biochim. Biophys. Acta* **721**: 349-356

Legerton, T.L., Kanamori, K., Weiss, R.L., and Roberts, J.D. (1983) Measurements of cytoplasmic and vacuolar pH in *Neurospora crassa* using nitrogen-15 nuclear magnetic resonance spectroscopy. *Biochem.* **22**: 899-903

Lenz, A.-G., and Holzer, H. (1984) Effects of chloroquine on proteolytic processes and energy metabolism in yeast. *Arch. Microbiol.* **137**: 104-108

Luvisetto, S., Schmehl, I., Cola, C., and Azzone, G.F. (1991a) Tracking proton flow during transition from anaerobiosis to steady state. 1. Response of matrix pH indicators. *Eur. J. Biochem.* **202:** 113-120

Luvisetto, S., Cola, C., Schmehl, I., and Azzone, G.F. (1991b) Tracking proton flow during transition from anaerobiosis to steady state. 2. Effect of cation uptake on the response of a hydrophobic membrane bound pH indicator. *Eur. J. Biochem.* **202:** 121-130

Maxfield, F.R. (1989) Measurement of vacuolar pH and cytoplasmic calcium in living cells using fluorescence microscopy. *Meth. Enzymol.* **173:** 745-771

Molenaar, D., Abee, T., and Konings, W.N. (1991) Continuous measurement of cytoplasmic pH in *Lactococcus lactis* with a fluorescent pH indicator. *Biochim. Biophys. Acta* **1115:** 75-83

Montrose, M.H., Friedrich, T., and Murer, H. (1987) Measurement of intracellular pH in single LLC-PK$_1$ cells: Recovery from an acid load via basolateral Na$^+$/H$^+$ exchange. *J. Membr. Biol.* **97:** 63-78

Nicolay, K., Scheffers, W.A., Bruineberg, P.M., and Kaptein, R. (1982) Phosphorous-31 nuclear magnetic resonance studies of intracellular pH, phosphate compartmentation and phosphate transport in yeasts. *Arch. Microbiol.* **133:** 83-98

Ohkuma, S., and Poole, B. (1978) Fluorescence probe measurement of the intralysosomal pH in living cells and the pertubation of pH by various agents. *Proc. Natl. Acad. Sci. USA* **75:** 3327-3331

Pilatus, U., and Techel, D. (1991) ^{31}P-NMR-studies on intracellular pH and metabolite concentrations in relation to the circadian rhythm, temperature and nutrition in *Neurospora crassa. Biochim. Biophys. Acta* **1091:** 349-355

Pontecorvo, G., Roper, J.A., Hemmons, L.J., MacDonald, K.D., and Bufton, A.W.J. (1953) The genetics of *Aspergillus nidulans. Adv. Genet.* **5:** 141-238

Pressman, B.C. (1976) Biological applications of ionophores. *Annu. Rev. Biochem.* **45:** 501-530

Rahim, A.T.M.A., Miyazaki, A., Morino, Y., and Horiuchi, S. (1991) Biochemical demonstration of endocytosis and subsequent resecretion of high-density lipoprotein by rat peritoneal macrophages. *Biochim. Biophys. Acta* **1082:** 195-203

Rashid, F., Horobin, R.W., and Williams, M.A. (1991) Predicting the behaviour and selectivity of fluorescent probes for lysosomes and related structures by means of structure-activity models. *Histochem. J.* **23:** 450-459

Rink, T.J., Tsien, R.Y., and Pozzan, T. (1982) Cytoplasmic pH and free Mg^{2+} in lymphocytes. *J. Cell Biol.* **95:** 189-196

Roos, A., and Boron, W. (1981) Intracellular pH. *Physiol. Rev.* **61:** 296-434

Roos, W., and Slavík, J. (1987) Intracellular pH topography of *Penicillium cyclopium* protoplasts. Maintenance of pH by both passive and active mechanisms. *Biochim. Biophys. Acta* **899:** 67-75

Seksek, O., Henry-Toulmé, N., Sureau, F., and Bolard, J. (1991) SNARF-1 as intracellular pH indicator in laser microspectrofluorometry: A critical assessment. *Anal. Biochem.* **193:** 49-54

Slavík, J. (1982) Intracellular pH of yeast cells measured with fluorescent probes. *FEBS Lett.* **140:** 22-26

Slavík, J. (1983) Intracellular pH topography: determination by a fluorescent probe. *FEBS Lett.* **156:** 227-230

Slayman, C.L., and Zuckier, G.N. (1989) Proton-potassium symport in walled eukaryotes: *Neurospora. Meth. Enzym.* **174:** 654-667

Thomas, J.A., Buchsbaum, R.N., Zimniak, A., and Racker, E. (1979) Intracellular pH measurements in Ehrlich ascites tumor cells utilizing spectroscopic probes generated *in situ. Biochemistry* **18:** 2210-2218

Tsujimoto, K., Semadeni, M., Huflejt, M., and Packer, L. (1988) Intracellular pH of halobacteria can be determined by the fluorescent dye 2',7'-bis(carboxylethyl)-5(6)-carboxyfluorescein. *Biochem. Biophys. Res. Commun.* **155:** 123-129

VERY LOW LEVEL FLUORESCENCE DETECTION AND IMAGING USING A LONG EXPOSURE CHARGE COUPLED DEVICE SYSTEM

M.H.F. Wilkinson, G.J. Jansen, and D. van der Waaij

Laboratory for Medical Microbiology, University of Groningen, Oostersingel 59, 9713 EZ Groningen, The Netherlands

SUMMARY

A system to improve the sensitivity of an industrial charge coupled device (CCD) video camera by increasing the exposure time is presented. The system consists of an expansion board for the IBM-PC, and interface software. The gain in sensitivity is shown to be a linear function of exposure time. Exposures of up to 8.5 seconds have been tested and they show that a 256-fold increase in sensitivity is attainable.

INTRODUCTION

In our laboratory an automated system for quantitative immunofluorescence measurement and imaging has been developed (Apperloo-Renkema *et al.*, 1991). The main problem in immunofluorescence imaging is the low level of light. Two types of camera have adequate performance: cooled CCD cameras, and image intensifier cameras. The latter type is more expensive than the former, and is rarely suitable for normal light levels, which is necessary in our system, since we need to combine phase contrast and fluorescence images from the same camera. An alternative is to use special purpose cooled CCD systems, designed specifically for these situations. These too are expensive. Our image processing systems all use unmodified, industrial CCD-cameras which have frame rates of 30 frames per second, and hence a 33 millisecond exposure time in their normal operating mode. Using a special feature of these cameras, we have now improved their sensitivity by

Biotechnology Applications of Microinjection, Microscopic Imaging, and Fluorescence, Edited by P.H. Bach *et al.*, Plenum Press, New York, 1993

simply increasing the exposure time. Control over the exposure time is provided through an expansion board for the IBM-PC developed in our laboratory.

MATERIALS AND METHODS

The Image Processing Systems

Our image processing systems consist of 80386 or 80286 based IBM-PC/AT compatible computers, each equipped with a MATROX MVP-AT frame grabber and image processing board (MATROX Electronic Systems Ltd, Dorval, Quebec), and a Fairchild CCD 5000/1 camera (Fairchild Weston Systems Inc., Sunnyvale, California, USA). The frame grabber of the MVP-AT boards can capture images from standard American and European format video cameras (RS170A and CCIR formats, respectively). They have no input facilities for other devices, such as special purpose CCD systems. The Fairchild cameras comply with the RS170A format and have a special purpose connector that allows access to and manipulation of all important video timing signals. The software consists of a number of programs for image acquisition and analysis, all of them developed in our laboratory (Meijer *et al.*, 1990; Apperloo-Renkema *et al.*, 1991; Meijer, 1991), with the aid of the IMAGER-AT libraries of functions, supplied by MATROX. The programming languages used are Microsoft Pascal 4.0 and C 5.1 and occasionally assembly (Microsoft Macro Assembler). In this experiment a Leitz Orthoplan fluorescence microscope, equipped with a HBO-100 mercury vapour lamp (Osram), was used.

The Expansion Board

An experimental expansion board has been developed that provides an interface between the camera's special purpose connector and the host computer. The board has been built from standard logic chips from the low power Shottky transistor-transistor logic (LS-TTL) family (Texas Instruments, Dallas, Texas, USA, and Philips, Eindhoven, Netherlands). Chips from other TTL families could also be used, since there are no critical timing problems (Texas Instruments, 1987). The circuit has been built on a prototype board, which has an IBM-PC connector on one side, and rows of holes 0.1 inch apart to mount and solder the components. These boards can be obtained in most stores that supply electronic components. Because the board contains only 11 chips and a handful of other components, the design was done by hand. Data and many useful tips on interfacing electronics to the IBM-PC have been compiled by Sargent and Shoemaker (1986). The board allows access to a number of video signals, some of which are first processed. The board implements a read-only port which allows access to the vertical drive (VD), frame index (FI) and frame clock (FCK) timing signals. The first is a signal supplied by the camera, and is passed unmodified. A pulse appears twice every video frame on this line. The second is supplied by the camera, but the pulse's duration is extended by the board, from 68 nanoseconds to 1 millisecond, to facilitate reading. FI pulses appear at the beginning of each odd frame half. These first two signals are combined on the board to form FCK, which indicates whether the video camera is in an odd or even frame-half. The vital timing signal is the charge transfer control (XFER). Two write-only ports allow the computer to set or reset this timing signal. The effect of setting the XFER is to block the transfer of charge from the CCD chip at the heart of the camera. This means that the CCD's "shutter" remains effectively open, without obstructing other video timing signals. A completely blank, but otherwise normal video signal is still transmitted to the frame grabber during exposure. This is vital, since the frame grabber would otherwise not be able to record the exposed image properly. Resetting XFER transfers all of this charge to the video-image circuitry at once, briefly producing a long exposure image on the video screen. The board also has an extra read-only port which allows software to detect the presence of the board automatically. Reading from this port always yields the value of 240. The port addresses can be altered to eight different values, using jumpers, to avoid conflicts with other hardware. The board also uses the RESET line on the computer's bus, to reset XFER whenever the computer is reset. This means that the board is "transparent" to programmes that do not use it. In other words, as long as nobody uses the board, the cameras behave as they have always done. After each long exposure the camera is automatically reset to normal operation. The electronic circuit design is available on request. Details of the camera operation can be found in the camera's manual and in databooks supplied by the manufacturer (Fairchild Weston, 1987).

The Software

The software's task is to synchronize the operation of the frame grabber board and our exposure control board. One procedure, dubbed LONGEXPOSURE, performs this task. It first sets the frame grabber into continous frame grabbing mode, waits until the frame clock (FCK) is HIGH, sets XFER, waits for the required number of pulses on the vertical drive (VD) line, and resets XFER when the FCK is high again. It then waits for one more VD pulse and instructs the frame grabber to stop its continuous grabbing mode. The result is a long exposure image, stored in the frame grabber's memory. This procedure has been inserted into the earlier programmes used for immunofluorescence work (Apperloo-Renkema et al., 1991). The exposure time is user-selectable to any positive integer number of frames. The programmes automatically detect the presence of the board using the "board present" port, checking which of the eight possible addresses gives the correct response. Digital image processing techniques used in our software can be found in Gonzales and Wintz (1987), Tou and Gonzales (1974), and Lindley (1991). The programming techniques, hierarchy and morphological applications of our image processing system are described by Meijer (1991) and Meijer et al. (1992). All programmes run under the MS-DOS (or compatible) operating system.

The Specimens

1. Immunofluorescent. As a first test of the system, immunofluorescent slides of a sample of faecal bacteria and of a *Klebsiella* strain where used. For specimens of faecal bacteria, 0.5 g of faeces was suspended in 4.5 ml Tween-80 (0.5%) and homogenized on a Vortex mixer for 2 minutes. The faecal suspension was subsequently centrifuged for 10 minutes at 1,490 x g in a Beckman centrifuge type TJ-6 (Palo Alto, California, U.S.A.). Slides (Immunocor, France) were degreased for 10 minutes in acetone and were coated with a 10% poly-L-lysine solution (Sigma Diagnostics, St. Louis, U.S.A) for 10 minutes in order to ensure proper adhesion of the faecal bacteria to the microscope slide. Ten µl of the faecal supernatant was pipetted in each of the 12 wells per slide. After 10 minutes of fixation in acetone each well was incubated with 20 µl of a matching serum dilution (10% serum in pH 7.2 phosphate buffered saline (PBS) and a temperature of $37^{o}C$) during 45 minutes at $21^{o}C$ in a moist chamber. The slides were dried and washed three times, each for 5 minutes with PBS of $37^{o}C$. After drying, each well was incubated with 20 µl fluorescein isothiocyanate

Figure 1. Immunofluorescence image of faecal bacteria, 4 frames (0.133 seconds) exposure.

(FITC)-conjugated goat anti-human F(ab′)2 IgA, IgM and IgG (Kallestad, Texas, U.S.A., 1% conjugate in pH 7.2, PBS, 37°C) for 1 hour at 21°C in a moist chamber. Finally, the slides were washed three times for 5 minutes in PBS of 37°C. Mounting fluid, consisting of pH 8.6 glycerol/Tris buffer (0.1 M) 1:1, was added to the slides before they were covered with a cover slide. The *Klebsiella* strain was cultured in brain heart infusion broth (BHI, Oxoid, Basingstoke, UK). One ml of the resulting suspension was centrifuged in a Sigma 201 M centrifuge at 11,430 x g for 15 minutes. The supernatant was discarded and the pellet was resuspended in demineralized water with 0.5% Tween-80. This suspension was treated in the same way as 4.5 ml faecal suspension. A number of immunofluorescent slides of skin tissue, including negative controls, were obtained from the Department of Dermatology, University of Groningen. All these specimens used FITC-labelled conjugates. Various exposures of these specimens were made to assess the effect of the exposure time on sensitivity and image quality qualitatively.

2. Acridine Orange. To assess the quantitative gain in system sensitivity, Acridine Orange stained *Escherichia coli* were used, because this stain did not fade under ultraviolet exposures lasting several minutes, i.e there is no photobleaching. The *E. coli* were fixed following the procedure described above. The slide was then stained with Acridine Orange using the method described by McCarthy and Senne (1980). A sequence of images was

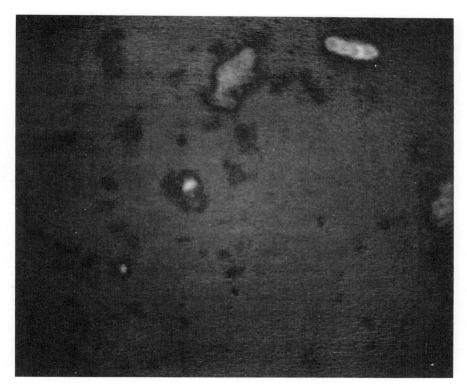

Figure 2. Immunofluorescence image of same slide as in Figure 1, but 60 frames (2 seconds) exposure.

recorded with exposure times of 8, 16, 32, 64, and 128 frames (approximately 0.25 to 4.2 seconds). Longer exposure times caused camera saturation on these relatively bright specimens. The fluorescence of three species of unstained bacteria (*E. coli*, *Staphylococcus epidermidis*, and *Streptococcus* sp.) was measured at exposure times of 8, 16, 32, 64, 128 and 256 frames (0.25 to 8.5 seconds), as a negative control.

Analysis

All images of the immunofluorescence specimens except for the brightest were displayed in pseudocolours to enhance contrast. All photographs displayed were taken directly from the screen and reproduced in black and white.

Quantitative measurements were done on the Acridine-Orange fluorescence using the procedure described by Apperloo-Renkema *et al.* (1991). This method compares the average intensity of the pixels within the object with the average intensity of the pixels in the immediate surroundings of the object. The shape of each object is determined by a phase contrast image of the same field of view, taken with the same camera, but operating in its normal mode. Dust in the optical path is corrected for by subtracting an image of an empty, or defocussed, field of view from the phase contrast image. Objects occurring in

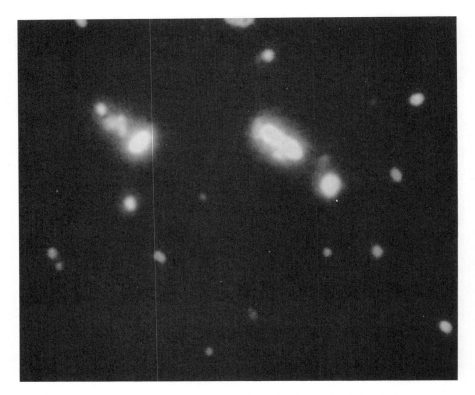

Figure 3. Immunofluorescence image of *Klebsiella* strain isolated from sepsis patient, 60 frames (2 seconds) exposure.

both background image and bacterial image are thus eliminated. Detection of objects is done automatically, using an automatic threshold selection method described by Kittler *et al.* (1985). After smoothing the image using a 3 x 3 averaging filter (e.g. Lindley, 1991), a weighted average of the pixel-values is computed for each area of the image. The gradient at each point is used as weight. The resulting weighted average is used as threshold: all pixels darker than this threshold are considered part of an object, all lighter pixels not. The user can delete obvious artifacts (e.g. fibres or crystals) manually by pointing at them with a mouse, though this is rarely necessary. For each exposure time, some 300 objects were measured, and the median of the fluorescence of these objects was taken as indicative of the system's sensitivity, as function of exposure time. Standard linear regression of median fluorescence level versus exposure time was done on the results.

RESULTS

Immunofluorescence

Figures 1 and 2 compare the image quality of short and long exposure shots (4 frames and 60 frames, respectively) using the same camera, on the same slide of faecal

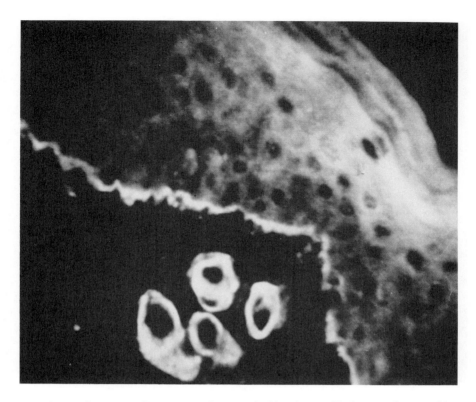

Figure 4. Immunofluorescence image of skin tissue, 90 frames (3 seconds) exposure.

bacteria. Different fields of view were used to avoid problems with photobleaching. Visually, the fluorescence of both fields of view seemed similar. The increase in quality and sensitivity in the long exposure shot is evident. Very little fluorescence could be seen with the naked eye. Similar results where obtained with the *Klebsiella* strain and the specimens of skin tissue as can be seen in Figures 3, 4 and 5. The negative control (Figure 5) shows a faint but distinct image. Long exposure shots (more than 3 to 4 seconds) showed some signs of a background fog building up. This is probably due to thermal noise in the CCD chip.

Acridine Orange

Figure 6 shows the median brightness of the Acridine Orange stained *E. coli* as a function of exposure time, along with the negative controls. For the Acridine Orange stained bacteria a clear linear relationship is visible. The coefficient of correlation is good (0.984), and the intercept is not significantly different from zero. The *E. coli* negative control group also shows a consistently rising linear trend, with a coefficient of correlation of 0.877. The *Streptococcus* and *Staphylococcus* sp. give similar results, with similar slopes, and even better coefficients of correlation (0.993 and 0.978, respectively). All slopes are significantly positive at the 99% level of confidence. No fluorescence could be seen with the naked eye in these control slides.

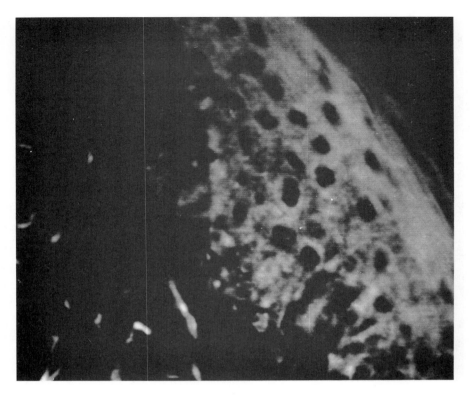

Figure 5. Immunofluorescence image of skin tissue, negative control, 120 frames (4 seconds) exposure.

DISCUSSION

All results indicate that a significant increase in sensitivity can be reached by increasing the exposure time. The apparent "fogging" of the background is probably due to thermal noise in the CCD, and could be reduced by such common image processing algorithms as background subtraction and smoothing (e.g. Gonzales and Wintz, 1987). One way of acquiring a good background image for subtraction could be to shut the camera off from all light and record an image with the same exposure time as the original image. Subtracting the background from the original should reduce the fogging. Further cooling of the CCD by means of an external cooling unit is another option that should be investigated. As expected, the increase in sensitivity is linear for stains that do not show photobleaching. Since the longest exposure used (256 frames = 8.53 seconds) is 256 times the normal exposure, it is safe to say that the sensitivity must have increased by the same factor. The consistently rising trend in the negative control groups could well be explained by autofluorescence. If this is the case, the conclusion must be that the bacteria themselves limit the sensitivity of immunofluorescence, not the sensitivity of our cameras. The fact that all negative controls measured so far showed a small positive signal supports this conclusion. The system does have its limitations of course. It cannot be used on rapidly varying signals, and thermal noise could well limit its performance, but the system performs

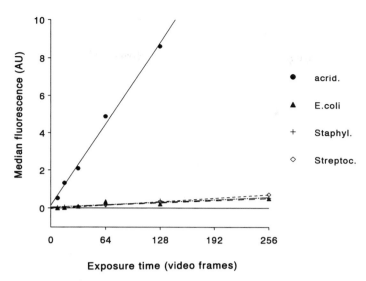

Figure 6. Fluorescence versus exposure time, for Acridine Orange stained *E.coli* and unstained *E. coli*, *Staphylococcus epidermidis* and a *Streptococcus* species.

very well when a combination of bright-field or phase contrast and fluorescent images is needed. As far as the construction of the expansion boards is concerned, building on prototype boards has its limitations. At this moment two prototype boards are performing well in two of our five workstations, but to equip all of them with prototypes would be courting disaster. Soldering all the elements together is messy and error prone, so we are designing a printed circuit edition of the board. Once this has been done making small series of boards will be very much easier and safer. The portability of the software may be a problem, since interruptions by other programmes in multi-tasking environments (e.g. MS-Windows or UNIX) could cause timing problems. Under MS-DOS no such problems exist, unless ill-behaved terminate-stay-resident programmes are present. Under MS-Windows, options do exist to claim the whole computer for a while, so it should be possible to use this software in that environment, but under UNIX a different approach is probably needed, such as linking a device driver to the operating system kernel.

CONCLUSION

The system described boosts the sensitivity of a standard industrial camera to a level comparable with many special purpose CCD systems. It cannot match the sensitivity of image intensifiers or photon counters, but the evidence strongly suggests that for our work it does not have to, as we consistently find a slight positive signal in our negative controls. The great advantage of a system with variable exposure is the ease with which it can be adapted to greatly varying light levels. Combining phase contrast and immunofluorescent images is just one such instance. The obvious disadvantage is that a long exposure system cannot respond rapidly: real-time video of faint images is not feasible. If one is only interested in slowly varying signals, and needs to compare the fluorescent (or luminescent) image with a phase contrast (or bright field) image, CCD systems have a clear advantage. Finally, the advantage of this particular system is its modest price compared to other systems, and the fact that it can be adapted to normal video processing equipment, such as frame grabbers.

ACKNOWLEDGEMENTS

We would like to thank Dr H.Z. Apperloo-Renkema for her slides of faecal bacteria, and Drs S. Bruins and M.C.J.M. de Jong of the Department of Dermatology for their slides of skin tissue. This research is sponsored by SymbioPharm GmbH and Co. KG, Herborn, Germany. Finally we would like to thank the European Commission for supporting this Workshop.

REFERENCES

Apperloo-Renkema, H.Z., Wilkinson, M.H.F, Oenema, D.G., Van der Waaij, D. (1991) Objective quantitation of serum antibody titres against enterobacteriaceae using indirect immunofluorescence, read by videocamera and image processing system. *Med. Microbiol. Immunol.* **80**: 93-100

Fairchild Weston (1987) "1987 CCD Databook", Fairchild Weston, Sunnyvale, California, 185-191

Gonzales, R.C., and Wintz, P. (1987) "Digital Image Processing", Addison-Wesley, Reading, Massachusetts, 139-388

Kittler, J., Illingworth, J., and Föglein, J. (1985) Threshold selection based on a simple image statistic. *Comp. Vision Graph. Image Proc.* **30**: 125-147

Lindley, C.A. (1991) "Practical Image Processing in C", Wiley, New York, 347-382

McCarthy, L.R., and Senne, J.E. (1980) Evaluation of Acridine Orange stain for detection of microorganisms in blood cultures. *J. Clin. Microbiol.* **11**: 281-285

Meijer, B.C., Kootstra, G.J., and Wilkinson, M.H.F. (1990) A theoretical and practical investigation into the characterization of bacterial species by image analysis. *Binary Comp. Microbiol.* **2**: 21-31

Meijer, B.C. (1991) Medeyes, an extensible language for image analysis. *Binary Comp. Microbiol.* **3**: 57-63

Meijer, B.C., Kootstra, G.J., and Wilkinson, M.H.F. (1992) Morphometrical parameters of gut microflora in human volunteers. *Epidemiol. Infect.* (in press)

Sargent, M., and Shoemaker, R.L. (1986) "The IBM Personal Computer from the inside out", revised ed., Addison-Wesley, Reading, Massachusetts, 428-445

Texas Instruments (1987) "The TTL Data Book", Volume 1, Texas Instruments, Dallas, 2.2-3.701

Tou, J.T., and Gonzales, R.C. (1974) "Pattern Recognition Principles", Addison-Wesley, Reading, Massachusetts, 39-155

MODULATION OF NUCLEUS FORMATION IN MOUSE OOCYTES FUSED WITH EARLY AND LATE G2 BLASTOMERES

Cezary Wojcik, Ewa T. Mystkowska, Wojciech Sawicki, and Aldona Komar

Laboratory of Experimental Embryology, and Department of Histology and Embryology, Institute of Biostructure, Warsaw Medical Academy, ul. Chalubinskiego 5, 02-004 Warsaw, Poland

INTRODUCTION

The fusion of secondary oocytes and blastomeres is a good tool for the study of the cellular events which occur during early embryo development. The secondary oocytes contain the maturation promoting factor and, therefore, it might be anticipated that the fusion of oocytes and interphase blastomeres would produce premature condensation of chromosomes (Murray, 1989; McConell, 1990).

In the present investigation, we fused secondary oocytes with either early or late G2 blastomeres of the second cell cycle of mice, in order to find the fate of nuclei of fused cells. In the development of the mouse embryo, the transcription starts at the second cell cycle (Flach et al., 1982), whereas in the first cell cycle only the maternal mRNA is used for translation (for review, see Telford et al., 1990).

The G2 phase of the second cell cycle is exceptionally long and lasts approximately 12 hours (Sawicki et al., 1978), i.e. it occupies more than 50% of the entire cell cycle. At this phase of the cell cycle a substantial shift in molecular events occurs (Johnson and Pratt, 1983; McLachlin et al., 1983).

In the present study we tried to find the fate of nuclei in the cells formed by fusion from either early or late G2 blastomeres and secondary oocytes.

Biotechnology Applications of Microinjection, Microscopic Imaging, and Fluorescence, Edited by P.H. Bach *et al.*, Plenum Press, New York, 1993

231

MATERIALS AND METHODS

Oocytes and blastomeres were obtained from inbred CFW and F1 (CBl x CBA) mice after hormonal treatment.

Oocytes

Secondary oocytes were recovered 13 hours post human chorionic gonadotropin (hCG). Ovulated oocytes were released from the oviduct in hyaluronidase (200-300 IU/ml in phosphate buffered saline) due to the removal of corona radiata cells. The zona pellucida was digested by pronase, and zona-free oocytes were placed in M2 medium under paraffin oil at $37^{\circ}C$ (Fulton and Whittingham, 1978).

Blastomeres

Female mice were mated with the males of the same strain. Embryos were obtained from the oviducts either 38 (early G2 phase) or 48 hours (late G2 phase) post hCG. Two-cell embryos were deprived of zona pellucida with pronase, and their blastomeres isolated by multiple pipetting and placed in M2 medium under paraffin oil at $37^{\circ}C$.

Fusion Procedure

Oocytes and blastomeres were placed in M2 medium without BSA and transferred to M2 medium containing phytohemaglutinin (300 µg/ml). One oocyte became attached to one blastomere, and cell pairs were transferred into electrofusion chambers filled with isotonic glucose solution supplemented with Ca^{2+} and Mg^{2+}. The cell pairs were located between two platinum electrodes, separated by a distance of approximately 100 µm. Two pulses of 35-60 V voltage and 35 ms duration were applied (Kubiak and Tarkowski, 1985).Then the cells were transferred to M2 medium under paraffin oil at $37^{\circ}C$ and cell fusion was observed with a light microscope.

Cell Culture and Observation

Fused cells were transferred to M16 medium under paraffin oil at $37^{\circ}C$ and cultured for 0.5 - 1, 6, or 22-26 hours in 5% CO_2 (Whittingham, 1971). The cells were fixed in Heidenhein's fixative, whole mounted in egg albumin, and stained with Ehrlich's hematoxylin.

RESULTS

The G1 blastomeres were fused with each other and treated as the first control. The same was done for early G2, and late G2 blastomeres. All fused cells cultured for 24 hours divided and produced embryos. On the other hand, corona radiata-free and zona pellucida-free secondary oocytes received the electric shock and were cultured for 24 hours, as a second control. No signs of parthenogenetic activation were observed.

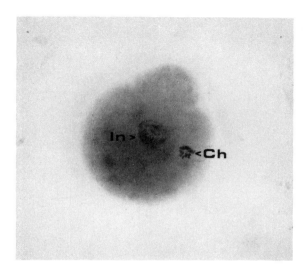

Figure 1. A cell that is being formed by the fusion of a secondary oocyte and a late G2 blastomere 15 minutes after electric shock. Interphase nucleus (In) of late G2 blastomere; chromosomes (Ch) of the metaphase plate of the secondary oocyte.

The fusion of secondary oocytes with blastomeres of either early or late G2 phase (Figure 1) led to premature chromosome condensation, which occurs as early as 0.5 hour after the electric shock. Then two separate chromosome groups were observed in most of the fused cells. However, 6 hours after the fusion, in most of the fused cells one chromosome group was observed; the number of chromosomes in such a group was higher than that in the chromosome groups observed 0.5 hour after the electric shock.

Table 1. The fate of fused cells as a result of electrofusion of early and late G2 blastomeres and secondary oocytes cultured for 24 hours, and expressed as numbers and percentages.

	Oocyte fused with:	
	early G2 blastomere	late G2 blastomere
Number and type of fused cells	40(100)[a]	32(100)
One cell		
Metaphase plate	13(33)	21(65)
Interphase nuclei		
one	7(18)	-
two or three	6(15)	-
Two cells with one or two nuclei	10(25)	2(6)
Fragmentation	4(10)	9(28)

[a] Figures in parenthesis represent percentage of the total number of fused cells.

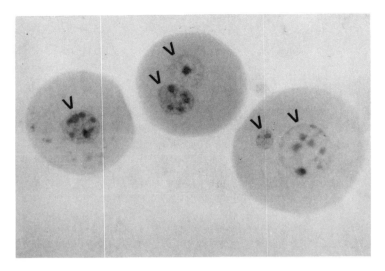

Figure 2. Three cells, each produced by the fusion of secondary oocyte with an early G2 blastomere, 24 hours after the fusion. Note in each cell 1-2 interphase nuclei (arrowheads) containing nucleoli.

Fused cells cultured for 22-26 hours produced the following structural patterns:-

a) one-cell embryoids,

b) two-cell embryoids containing pronucleus-like interphase nuclei, and

c) embryoids with fragmentation of the cytoplasm.

Among the one-cell embryoids, cells containing metaphase plate-like structures, and one or more interphase nuclei were observed. In some of the one-cell embryoids the micronuclei were also produced.

Table 1 presents the number and percentage of various forms of embryoids produced by the fusion of either early or late G2 blastomeres with secondary oocytes.

After the fusion of the secondary oocytes with early G2 blastomeres, the most frequent nuclear pattern was the interphase nuclei (Figure 2) that occurred in about 35% of fused cells. Approximately 25% of fused cells divided and in each of those cells 1-5 interphase nuclei were observed.

The fusion of the secondary oocytes with late G2 blastomeres produced metaphase plate-like structures in 65% of the fused cells (Table 1); some chromosomes could be dispersed in the cytoplasm (Figure 3). In this variant fusion product, the reconstruction of the interphase nucleus occurred in 6% of fused cells.

DISCUSSION

The secondary oocytes are blocked in metaphase of the second meiotic division and their fusion with either early or late G2 blastomeres of the second cell cycle leads to premature chromosome condensation of the chromatin of interphase nuclei (Rao and Johnson, 1974).

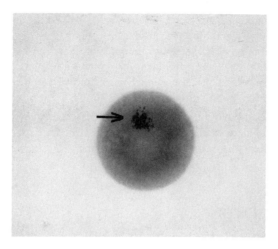

Figure 3. A cell produced by the fusion of a secondary oocyte and a late G2 blastomere, 24 hours after the fusion. Note one group of chromosomes (arrow) derived from fused cells.

In the present investigation we observed premature chromosome condensation half an hour and one hour post-fusion and the premature chromosomes endured at least for 6 hours post-fusion. Fusion of early G2 blastomeres with secondary oocytes yielded the interphase nuclei, whilst the fusion of late G2 blastomeres with oocytes failed to produce interphasal nuclei. It should be borne in mind that, in the course of the long G2 phase of the second cell cycle, the earliest signs of cell differentiation may occur.

The early structural and molecular events that appears as a consequence of fertilization are controlled by a post-transcriptional programme. This programme is, therefore, connected with the cytoplasm and has the features of maternal inheritance (Johnson *et al.*, 1984).

It has been pointed out (Braude *et al.*, 1979; Petzoldt *et al.*, 1980; Flach *et al.*, 1982; Bolton *et al.*, 1984; Smith *et al.*, 1990) that in the first cell cycle, the enucleation failed to modify the protein synthesis in cytoplasts. However, in the second cell cycle two surges of transcription have been detected: just before and soon after DNA synthesis. As a consequence of this transcription, most of the messenger molecules necessary for maintaining the third and fourth cell cycles appear (Johnson and Pratt, 1983; McLachlin *et al.*, 1983). All these data provided the evidence that early mouse embryo development is controlled by cytoplasmic cell cycle regulators.

In our study, the blastomeres of early G2 phase represent the stage of embryo development at which the transcription and the translation of the regulatory molecules has not yet started. On the other hand the blastomeres of late G2 phase could have completed their synthesis of molecules.

It can be speculated that the cytoplasm of early G2 blastomeres either contains the factor(s) for nuclear reconstruction or do not have the blocking molecules which prevent nuclear reconstruction. The latter may be present in the cytoplasm of late G2 blastomeres and, therefore, prevent nuclear reconstruction.

Whatever the nature of the factor promoting or preventing nuclear reconstruction, the reconstructed nucleus resembles the pronucleus rather than fully formed interphase nucleus. It is tempting to speculate that the pronucleus-like structure of the reconstructed nucleus is due to A/C lamins of the cytoplasm of secondary oocytes (Kubiak *et al.*, 1991). Lamins A/C are nucleoskeletal proteins, which were shown to correlate with pronuclear appearance of nuclei, unlike lamins B which are ubiquitous.

As far as the modification of the chromatin of fused cells is concerned, it should be mentioned that the decondensation of chromosomes and their participation in the nuclear reconstruction may reflect the activation of the oocytes. This, however, requires further study.

ACKNOWLEDGMENTS

We acknowledge the WHO Small Supplies Programme. One of us (C.W.) is deeply grateful to Noran International - Tracor Europa, U.K., Ltd., for supplying financial support, which made participation in this Workshop possible.

REFERENCES

Bolton, V.N., Oades, P.J., and Johnson, M.H. (1984) The relationship between cleavage, DNA replication and gene expression in the mouse 2-cell embryo. *J. Embryol. Exp. Morph.* **79**: 139-163

Braude, P.R., Pelham, H., Flach, G., and Lobatto, R. (1979) Post-translational control in the early mouse embryo. *Nature* **282**: 102-105

Flach, G., Johnson, M.H., Braude, P, R., Taylor, R.A.S., and Bolton, V.N. (1982) The transition from maternal to embryonic control in the 2-cell mouse embryo. *EMBO J.* **1**: 681-686

Fulton, B.P., and Whittingham, D.G. (1978) Activation of mammalian oocytes by intracellular injection of calcium. *Nature* **273**: 149-151

Johnson, M.H., McConell, J., and Van Blerkom, J. (1984) Programmed development in the mouse embryo. *J. Embryol. Exp. Morph.* **83**: 197-231

Johnson, M.H., and Pratt, H.P.M. (1983) Cytoplasmic localizations and cell interactions in the formation of the mouse blastocyst, *in:* "Time, Space and Pattern in Embryonic Development", W.R. Jeffrey and R.A. Raff, eds., Alan R. Liss Inc., New York, 117-128

Kubiak, J.Z., Prather, R.S., Maul, G.G., and Schatten, G. (1991) Cytoplasmic modification of the nuclear lamina during pronuclear-like transformation of mouse blastomere nuclei. *Mech. Develop.* **35**: 103-111

Kubiak, J.Z., and Tarkowski, A.K. (1985) Electrofusion of mouse blastomeres. *J. Embryol. exp. Morph.* **18**: 155-180

Murray, A.W. (1989) Cyclin synthesis and degradation in the embryonic cell cycle. *J. Cell Sci. Suppl.* **12**: 65-76

McConell, J. (1990) Cell cycle control in early mouse development. *J. Reprod. Fertil.* **42**: 205-213

McLachlin, J.R., Cavaney, S., and Kidder, G.M. (1983) Control of gap junction formation in early mouse embryos. *Devl. Biol.* **98**: 155-164

Petzoldt, U., Hoppe, P.C., and Illmensee, K. (1980) Protein synthesis in enucleated fertilized and unfertilized mouse eggs. *Wilhelm Roux' Arch. Devl. Biol.* **189**: 215-219

Rao, P.N., and Johnson, R.T. (1974) Regulation of cell cycle in hybrid cells. Cold Spring Harbor Conf. *Cell Proliferation* **1**: 785-800

Sawicki, W., Abramczuk, J., and Blaton, O. (1978) DNA synthesis in the second and third cell cycles of the mouse preimplantation development. *Exp. Cell Res.* **112**: 199-205

Smith, L.C., Wilmut, I., and West, J.D. (1990) Control of first cleavage in single-cell reconstituted mouse embryos. *J. Reprod. Fert.* **88:** 655-663

Telford, N.A., Watson, A.J., and Schultz, G.A. (1990) Transition from maternal to embryonic control in early mammalian development: a comparison of several species. *Mol. Reprod. Dev.* **26:** 90-100

Whittingham, D.C. (1971) Culture of mouse ova. *J. Reprod. Fert.* **14:** 7-21

INDEX